The Art & Adventure of BEEKEEPING

ORMOND & HARRY AEBI

ILLUSTRATED BY ERIC MATHES

1975

UNITY PRESS
SANTA CRUZ

PUBLISHED AUGUST 1975 BY
UNITY PRESS, BOX 1037, SANTA CRUZ
CALIFORNIA
95061

DESIGN & CALLIGRAPHY
BY
ERIC MATHES

LIBRARY OF CONGRESS CATALOGING IN PUBLICATION DATA

AEBI, ORMOND, 1916 -
THE ART & ADVENTURE OF BEEKEEPING
1. BEE CULTURE I. TITLE

SF 523.A29 638'.1 74-14661
ISBN 0-913300-39-X
ISBN 0-913300-38-1 (PBK)

PRINTED IN
THE
UNITED STATES OF AMERICA

To our friend Gary Blankenbiller, who has a most delightfully severe case of "Bee Fever" and who has been of great assistance and inspiration, this book is gratefully dedicated.

Introduction

My father and I have been keenly interested in honeybees all of our lives as were our ancestors before us. Our goal has been to set a new world's record for the production of wild flower honey from one hive in one season. To that end we have spent much time in observation, research, and experimentation with honeybees. Included with our personal adventures is a manual of instruction for the aspiring beekeeper and for all others who are students of nature. The purpose is twofold: To encourage many people to keep a few hives of bees; then show them how to be successful in obtaining the maximum amount of honey per hive.

The book has four primary objectives: To give information; to dispel fear; to give suggestions on how to encourage bees to make surplus honey for their keeper; to give each reader a personality examination so that he may personally evaluate his chances of becoming a successful beekeeper.

Bees are lovable little creatures but they are set in their ways and a beekeeper must be able to quickly adjust his thoughts and actions to their changing moods and manners as the season advances. He cannot readily change his bees. It is he who must make the required adjustments. An attempt has been made to illustrate this necessary type of thinking by the way in which many chapters have been written. More beekeepers are needed to help our friends the honeybees gather larger quantities of the abundant honey available every spring and summer that presently goes largely to waste.

We wish to express our heartfelt gratitude to Alice Shannon, LaVerne Faris, Margaret Haun, Aline Kinread, Mary Louise Goldman, Rita Tarrico, Darleen Johansen, Joan Minnick, Mathilde Sivertsen, Elizabeth Blackmun, Rod Westlund, Gilbert Silva, Thomas Platner, Gaylord Snavely, Fred Stoes, Dr. Donald Harner, and Dr. Harold Sundean.

We also wish to thank Walter Hard, Jr., Editor of Garden Way Publishing Company for his many encouraging letters, excellent critique, and helpful suggestions in the preparation of this work. And to all other friends and beekeepers who have expressed their abounding faith in the merits and usefulness of this book, our grateful appreciation and thanks.

ILLUSTRATIONS

TABLE OF CONTENTS

1

Early Adventures With Bees

BEE FEVER! Bee fever? The old-timers used to call it that when a man fell in love with honeybees, took the plunge and acquired his first hive. Some women get this blessed malady as joyfully as men. According to that definition I have always had bee fever and my father and mother had it before me for I cannot remember when we did not have the blessing of at least a few hives of bees near our house or out in the woods or hidden somewhere. If possible, we have had four or five within fifty feet of the back door of our house so that we might observe them every time we stepped out. Mother was glad her men had bee fever, for when we were out of her sight she always knew where we were—we were somewhere watching the bees. As soon as her housework was finished, Mother came outside to find us and together we studied our industrious little friends. She could often spot a swarm clustered on a hidden limb before Father or I had found it and was at all times on the alert for passing swarms of bees.

I was born in that part of Oregon largely covered with forests of Douglas fir and White oak. The honeybees could not work on the fir trees or the oaks but they did get a large amount of nectar, the sweet liquid raw material from which bees make honey, from the vast growths of poison oak bushes and vines that always seem to grow near oak trees. Poison oak blooms in the springtime with clusters of tiny whitish blossoms so small that one would hardly notice them were it not for the honeybees buzzing around and on them. Poison oak honey? Yes, indeed. And a fine flavor it has, too. It is perfectly safe to eat, as bees never get poison oak themselves nor do they ever carry it to their hives. Poison oak honey was my introduction to delicious honey and the marvelous world of the honeybee.

As a small boy I loved to watch the honeybees as they were hard at work collecting nectar and pollen. Bees mix nectar and pollen into a kind of dough and use it as food for their baby bees. It is a rich food so the baby bees (brood) grow very rapidly. Since there are thousands and thousands of little ones in each hive in the springtime, the adult field bees must be nimble and clever as they climb from one blossom to the next on the clusters of poison oak bloom.

In order to see the bees more clearly, I pulled down the poison oak vines. The bees always seemed to be in a hurry to collect their loads of nectar or pollen and never paid any attention to me. Before long I came down with a severe case of poisoning, my body being covered from head to toe with a weeping, itching rash that would not heal. Doctors were few and far away in the last days of the horse and buggy; besides, they cost money and money was at that time "scarce as hen's teeth," as my father used to say.

I sat in misery and cried. Mother tried every remedy she knew but none gave me any relief. At last one morning she decided to telephone long distance to Grandmother who might know a cure. There was no charge for that particular long distance telephone call in those days. However, Mother had to wait until all lines were clear the entire distance before a call could go through and that might be from ten minutes to three days—too often the latter. There was no way to tell when the distant lines would be clear, so Mother had to keep trying. She got her call through to Grandmother in about ten hours. This was surprisingly soon as we were on a seventeen-party line and some of the farther lines had as many as twenty subscribers. As the telephone was an easily accessible news media, everyone "rubbered" when the telephone rang. Rubbernecks lowered the electrical power in the telephone lines forcing Mother to shout her questions into the mouthpiece and Grandmother to yell on her end of the line in reply.

Everyone along the line knew that little Ormond had poison oak and heard Grandmother's cure. Before she hung up Mother told Grandmother that she would phone again in three days sometime around noon to let her know if I was well. This alerted everyone to keep off the lines and listen for Grandmother's three long rings and two short ones as Central relayed Mother's call to Grandmother's hand-cranked battery-powered wall telephone.

Grandmother's cure for poison oak was simple and easily obtainable.

"Just take the axe," she said, "and go to the nearest oak tree. Chop off a half milk pail full (about ten pounds) of oak bark. Fill the copper wash boiler about half full of water, dump in the oak bark, and boil it for an hour. After it has cooled somewhat, pour the dark brown water, bark and all, into the bathtub (it was the family washtub for clothes in those days) and give little Ormond a bath. Let him sit in the tub until the water becomes cool. In the meantime take a dipper and pour some of the brown water all over him—in his eyes, ears and everywhere. Then tuck him in bed and in the morning he will be well."

I was! It worked! Praise God, I was well! Mother joyfully telephoned the good news to Grandmother and all the rubbernecks on the line rejoiced too. I still catch poison oak very readily but all through life this remedy has always worked for members of our family. Grandmother used to say, "Whatever ill in life there be, the remedy is nearby you see."

I should mention a word of caution when using this oak bark water. It will not stain skin but it will discolor clothing in a most unlovely manner and Mother found the stain impossible to remove.

My father knew many things about bees for our branch of the family has been beekeepers for generations. Father knew about keeping bees in box hives, barrels, gums, and eight and ten-frame hives. However, as the years passed he always wanted to know more about bees. During my childhood, I remember he began ordering government bulletins on beekeeping. When he found a particularly interesting passage, he read it aloud so that my mother, two sisters, and I could learn too. Soon we acquired enough knowledge and woodworking tools to build our own hives and supers with their necessary internal frames. This was a great step forward, for in addition to the bees my father had acquired before I was born, we now added new hives and put bees in them as we were able to catch wild swarms.

The hive body, the basic home area of the queen bee, drones, and

worker bees, resembles a covered wooden apple box with an opening in one end. In this box the worker bees build honeycombs wherein they store nectar and pollen. The queen also lays eggs in part of these combs so that the bees may rear brood. The supers are boxlike compartments placed on top of the hive body when the bees need additional room to store surplus honey—honey for you and me.

My early memories of working with bees are very pleasant. Father placed a super on each hive about the middle of May. The strongest hives got another super near the middle of June. Around the last of August, on a nice warm afternoon, Father took off several supers and we had our first delicious fresh honey of the season.

We had no honey extractor in those days because it was an expensive item of equipment considering the number of hives we owned. The honey extractor was invented in Germany in the year 1865 and came to this country in 1867. This ingenious device soon revolutionized the whole beekeeping industry. Our present day extractors still use the same principle of centrifugal force to throw the honey out of the combs after uncapping without heat or pressure of any kind. When empty the combs may be replaced on the hives for the bees to refill. Bees value this arrangement as much as we do and are thus able to produce a great deal more honey each season.

However, as I was saying, we could not afford an extracting machine at that time so we had to use the age-old method of extraction. We cut the honeycomb out of the frames into a large clean container like a big dishpan and then, with a pair of thoroughly scrubbed hands and arms, one of us would reach in and crush the wax honey comb so that the honey could run out between our fingers. This was often my job. It was a race to see if I could crush up the combs as fast as my sisters cut them from the frames. It was a lot of fun to squeeze the sweet-smelling honeycombs and watch the honey drip off the ends of my fingers. Nevertheless, after an hour or so it became more like hard work. I would be "stickied up" with honey all the way to the elbows and so had to stay at my job, because if I removed my hands from over the honey pan, drops of honey and bits of wax fell on the floor—and there is nothing messier than a stickied up kitchen floor. When the honeycombs were all crushed, Mother poured the sweet gooey mass into a large colander that was placed over another big pan so the honey could drain away from the wax.

The wax was then removed, heated to the melting point, and made into cakes. We stored the honey in quart jars for future use or sale.

My mother was always thoughtful and observing. When she saw me getting tired, she would cut off a big bite-sized piece of honeycomb and put it in my mouth with a fork. How good that honeycomb tasted and how it did revive my spirits and renew my strength! Mother had the knack of knowing how to keep her little workers not only busy but also happy. I have learned since that bees predigest honey, and when eaten, it is taken directly into the bloodstream and therefore does give one a rapid pickup in energy. The happy effect of eating honey has been known for thousands of years. The Holy Bible speaks of it in First Samuel Chapter Fourteen and in many other places as well.

Honeybees were one of my chief delights in those days. I loved to watch them return to their hives loaded with nectar or pollen. It was easy to tell what they had been gathering. If they had nectar their abdomens would be extended beyond the length of their wings. I could calculate the size of the load by the length of the extension as a bee draws nectar into its honey sac located in its abdomen. Pollen is always carried in a little basket on each hind leg. Sometimes the bees came home with their baskets hugely filled, amounting in size to almost as large as a match head hanging from each hind leg. Then again on days when pollen was difficult to find, they came home with but very little in their baskets. Pollen comes in many colors from light bright yellow gathered from the plantain weeds to almost invisible dark gray from wild blackberry blossoms.

I do not remember being stung by a honeybee until I was twelve years old. I had been stung by yellow jackets on several occasions when I stepped on their nests in a hole in the ground. They were quick to sting in defense of their home. I had also bumped into a big hornets' nest made of gray paper-like material hanging from the limb of a tree. How they did sting! One day a bumblebee came into our kitchen and became entangled in the curtain by a window. I tried to free him and in doing so he became frightened and stung me. I'll never forget him!

We children always went barefoot in summer. One day while walking through some rank-growing marsh grass in a swampy place, a scorpion stung me on the foot just below the ankle. The agony of such a sting is almost beyond description. It was more than two months before the fiery burning pain finally ceased. This scorpion was a strange little creature about two and a half inches long, dark green in color and shaped somewhat like a crayfish or a lobster. At the extreme end of its tail it had a sharp poisonous point and it stung by striking with its tail. I have occasionally found a scorpion in a swampy place or under an old building where it is damp and dark.

Let me hasten to add that though the aforementioned creatures of God's creation did sting me, I in no way considered them my enemies but my friends. Yellow jackets will sting if we frighten them by stepping too near their nest hole or if we squeeze them while they are at work hollowing out a ripe plum or other ripe fruit. Nevertheless, they are friends of man for they catch and kill many houseflies and similar insects for their food. In the fall of the year when there are a great number of them, they will also sometimes attack hives of honeybees and kill some of the bees. In this case, close observation will show that the yellow jackets, as a rule, kill only the crippled, weak, or sick bees. They kill the bee, then cut it in two at the narrow area (petiole) between the thorax and the abdomen, leaving the thorax and carrying the abdomen to their nests. This is good. Killing the unfit bees makes for a strong healthy hive for wintering. In rare instances a large number of yellow jackets attack a single hive. In this emergency we must help or our bees are doomed. Immediately, we begin to search for the yellow jacket's nest in the ground.

In my experience the nest, or perhaps several of them, will be found within a 200-foot radius of the beehive. They are seldom found in open ground but nest in a sheltered place—in some bushes, near a fence post or light pole, a low brush heap, fence row, or near the top of a ditch bank. If the yellow jackets are killing our bees in great numbers, they must in return be killed. To do this, we get about two gallons of used crankcase oil or any other motor oil. When evening comes, we go to the nest and pour this oil all over and around the entrance. Next we take a spade and jam it into the ground many times all around and through the nest, cutting it completely to pieces. The yellow jackets become entangled in the oily earth and die. If we are quick and positive in all our actions, we do not get stung and our hive is saved.

Hornets are also our friends. They, too, catch and kill many houseflies and other larger flies and insects.

Bumblebees are not insect eaters but gather honey in the same way as honeybees, only they work alone. Their nest is in a small hole they make in the ground—usually on the open sunny slope of a hill-side. Six or eight inches below ground level they enlarge the hole to make space to rear their brood and store a thimbleful of honey for food. We often observed them on a warm hillside near the ocean south of San Francisco. I never had the heart to dig one out but those who did said the honey had an excellent flavor. Bumblebees pollinate red clover and other blossoms.

Just what role scorpions play in the ecology scheme I have been unable to discover but doubtless they, too, play a useful part. The one that stung me did so because I unknowingly stepped on its head. I could not blame it for stinging. I love all these little creatures, and the readers of this book will also as they become better acquainted with them.

Of all the insects I had met, the honeybees were my dearest friends. I loved them and they loved me until one day—well, it was a curious situation and I do not understand it even now. We had our bees located in a sunny ravine some distance from our house. One warm afternoon my father decided we should take off some honey. He would work with the bees and lift off the supers. After removing a super and smoking out the bees, he would bring it to me where I stood waiting for him under a shaggy old oak tree about one hundred feet away. When he came with a filled super it was my job to carry it the rest of the way to the house.

All went well. He had on his screen veil and gloves and was taking off his third super when I noticed a bee leaving the entrance to the hive. My tiny friend made a beeline straight for me. Surely she could not see me at that distance where I was standing under low-hanging branches in the shade of the big oak. With great interest I watched her come straight toward me true as an arrow. Too late I realized her evil intent. Zap! She struck me a hard stinging blow squarely between the eyes. I became acutely aware of how painful a bee sting can be. In a short time both eyes began to swell shut and continued to do so until I could not see except out of the far corner of each eye. That was a sad day but the beginning of an important discovery. The next time I got stung it did not swell up quite as much nor did it hurt as badly. We learned that not all stings are equal in intensity and that repeated stingings can cause one to become more or less immune to the poison.

Then came the financial crash of 1929 and the tragic depression that followed it. Our bees made honey—gallons of it. For the bees the seasons came and went as usual, the sun rose and set as always. There was an abundance of nectar to be gathered and they gathered it—making plenty of honey for themselves and a goodly surplus for us. But what could we do with the honey? We at least had an abundance of honey for food, but we could not eat it all and no one would buy it at any price. It was not that people suddenly disliked the taste of honey but rather that they had no money to pay for it. Money seemed to have disappeared from circulation. My Aunt Lydia had luscious ripe Royal Ann cherries to sell but no buyers, not even us, for we could not afford them either.

One day Father said to Aunt Lydia, "Lydia, how about trading a big bucket of cherries for a gallon honey?"

"I'll do that," said Aunt Lydia. This started an interesting chapter in our lives. Father bartered honey for almost anything we could think of and people were glad to trade. Thus our precious friends the honeybees had a real part in pulling us out of that quagmire stretch of history called the "Great Depression."

Somewhere, during those years, my father ran across an item discussing honey production wherein it was stated that a Mr. A. I. Root had in the 1890s set a world's record for production of wild flower honey from one hive in one season. It was the astonishing amount of 100 quarts or 300 pounds of wild flower honey from *one* hive in *one* season!

"Now there," said my father, "is the goal for us to shoot at. Let's go him one better and set a new world's record. How about that for an idea?" It was a challenging thought—the kind I love.

We took stock of where we were at that time. A very few of our hives made no surplus at all so we got nothing from them. Most hives made one surplus super or about twenty pounds of honey for us. Some made two supers or forty pounds. One or two, depending upon the season, made three supers or sixty pounds. We had thought that was a considerable quantity of honey. It really was, too, considering the location, the short bloom of wild flowers in that area, and our inefficient method of handling our bees.

We had begun to experiment with our bees and were having some degree of success in raising our production per hive when rumblings of World War II, already going on in Europe, began to be heard more loudly in the U.S.A. Late in 1939 we sold our ranch, bees and all, to our next-door neighbor and moved into the town of Dallas, Oregon. In October 1941 I was drafted into the Army—and my beekeeping days were at an end for the duration of World War II.

During the war there was never a dull moment. I worked with all my strength at whatever I was assigned to do, for I wanted to get the war over and go home. Wherever I was and whenever an opportunity came—and they were rare enough—I walked out into the countryside to look for bees, wild flowers, blooming trees and shrubs, and whatever had to do with bees in particular and wild life in general.

One such occasion I will never forget. For some months I had been working as typist in the Headquarters Office of Sheppard Field, Wichita Falls, Texas. I had been typing at top speed for many days, and for more than ten hours that day and was keyed up to a high pitch, when the officer in charge said to me, "Corporal Aebi, stop typing and take off somewhere before you fly to pieces." He really meant it so I rushed out of there and headed for the gate a half mile away and the wide open spaces beyond. It was late in the afternoon of a summer day. The reflected rays of the sun shone like a great sheet of flame low on the western horizon and black thunderheads were floating in the sky near at hand. But none of these things deterred me—neither did lack of supper—for I was free until morning.

I strode off down the first country road that led away from the Field and in a few hours was many miles away. Strangely enough, in all that distance I had not seen the flight of a single late-flying homegoing honeybee, nor had I passed even one farmhouse. Now that the first rush of nervous energy had worn off, I was both hungry and tired so I sat down on a ditch bank to rest. Night had fallen, but since the moon was at three-quarters full, everything near and far had a ghostly, fairy-like quality of exquisite beauty.

I love the night—always have. Back in the old days when we lived on the ranch, about eleven o'clock on a summer's night my little dog Queenie used to sit under my window and howl softly and expectantly. Soon I would hear my mother's voice from her bedroom next to mine, saying softly, "Ormond, are you awake?" And I would always answer, "Yes Mother." Then she would say, "Queenie's lonesome, why don't you take her for a run through the woods?"

What more could a boy ask? Away we would run over hill and hollow sometimes for miles. The full beauties and glories of such nights only the angels and fairies know but they let Queenie and me learn some of their secrets those nights. Queenie was black as a lump of coal—and wise. She never gave a bark or whine once we had left our own barnyard. This was vital to our pleasure and well-being for sometimes we came too near to a farmhouse or cabin away off at the other edge of the forest and the owner's dog barked and gave the alarm. Queenie never answered but came racing back to me and together we beat a hasty and silent retreat back into the big woods out of harm's way. Most people had "shootin' irons" in those days; so did my father, but no one ever took a potshot at Queenie or me. By one o'clock in the morning we would be back, panting and happy and much refreshed, and then to bed again.

It was a real joy to find myself away out in the lonely Texas countryside at night, this situation being to me not a fright, but a delight. The moon had risen higher in the sky, and I knew it was time for me to be heading back to the Field. The night had many voices, even out there on the dry plains of North Texas. Many sounds I could identify from old times as I walked along, but some were new to me and I wondered from what sources they might emanate. Then a huge cloud slowly drifted across the face of the moon and it became dark—pitch dark. No matter, I had already seen that the road lay straight ahead for a long way yet so I went on.

Suddenly I saw a strange sight. Over the fence to my left a pair of large luminous glowing blue eyes gazed steadily at me. Now this was something that I had never seen before. The eyes had an eerie glow. My hair almost stood up on end. The eyes did not move. Neither did I! But nothing happened so I took a few cautious steps down the road. There I saw another pair of big blue eyes staring at me. Another pause, another wait. Still nothing happened, so I moved on again walking backwards. At what seemed a safe distance, I turned around to walk forward again. But there they were—yet another pair of those big luminous blue eyes almost in the middle of the road and not more than forty feet away!

I gave up and stood still. Off to my left I noticed moonlight again bathing the earth with the light coming closer. The big cloud had almost passed the face of the moon. If I could just hold out a little longer. Then came the light—suddenly, brilliantly. All the big blue eyes disappeared and in their place was a flock of sheep. Just a flock of loving old woolly sheep, and one of them had squeezed through a hole in the fence and was standing in the middle of the road. I laughed aloud in my relief! Bless their dear old hearts, I had met this same flock of sheep on the way out and had been happy to give them my love and visit with them a few minutes. Whoever said that "Ignorance is bliss," must have been a most misinformed character, for I had not had such a fright in many a year. Just for the record —and I have owned sheep as well as bees for many years—it is seldom indeed that we see sheep with these glowing blue eyes. The darkness, coupled with reflected light, must have a certain degree of intensity before this phenomenon appears.

It had been quite an adventure and I realized that I had been as much afraid of those lovable ghostly sheep as most people are of my beloved honeybees. Seems all of us, given an unfamiliar or strange set of circumstances, are needlessly afraid of the unknown. Well, I thought, I will never be so easily frightened again.

The moon was brilliant for the moment so I hurried up the road toward the Field, now about three miles away. Its many lights cast a low reflection against the sky. Sudden darkness again, but I kept steadily on. Then I heard footsteps coming behind me. Footsteps behind me here at this time of night? Impossible. I hastened on but the footsteps gained on me slowly but steadily. The moon cleared for a moment. I whirled around. Nothing!

I walked on, descending a slight slope where the darkness became even more intense. The footsteps drew noticeably closer. I had heard some fishy tales about Texas but this really did "beat the racket" as my father was wont to say. I had been out this far once before so I knew that immediately ahead was a small culvert under the road with a low concrete abutment at each end. If I could just find this culvert, I would sit down on the low concrete wall at one end and wait for whatever was following me to pass by. To my great joy and relief I found it and sat down. The footsteps advanced—coming straight down the middle of the road—until they were abreast of where I sat. They paused for a moment and then came toward me and stopped right between my knees.

To say that I was downright puzzled would be to state the matter mildly. I could hear breathing, so it was at least a living creature and not a ghost. Moonlight suddenly flooded the landscape and there before me, looking up at me from between my knees, was the biggest and blackest, most beautiful skunk I had ever seen. I could have laughed aloud again in my relief. But I did not! Believe me, I knew better than to so much as breathe. Though he seemed like an old friend out of the past, for I had met his kind many times before, yet I knew better than to so much as move a muscle or wink an eyelid. His tail was up over his back as he looked steadfastly up at me. To those who know skunks this is the danger signal, indicating that one twitch or quiver will result in a squirt of sickeningly smelly burning yellow fluid right in the eyes. Queenie and I had the misfortune to learn it the hard way.

Mister skunk had an appraising look on his face, as though he had often wondered what one of those human creatures really looked like close up. For my part, I had never seen one of his kind quite so close to the end of my nose either. Then he raised his twitching nostrils almost up to mine, opened his red mouth with its gleaming white teeth, and smiled. Slowly, he backed out from between my knees and like an old gentleman of good breeding, he walked sedately on up the road. He was king in his own domain and he knew it. I followed at a discreet distance until, at the next sharp turn in the road, he went straight on across the fields as I headed for my barracks; arriving at one o'clock in the morning, tired and hungry but happy.

Here again I had learned another lesson of the wild. Old mister skunk had not squirted me with his repulsive protective fluid because I had neither frightened nor offended him. Later, I found out that as a rule honeybees will not sting either unless they are frightened or offended.

I was honorably discharged from the Army just in time to arrive home for Christmas 1945. But "going home" was no longer to go home to Dallas, Oregon, for during the war my father and mother had moved to San Francisco to work. My sisters had moved to faraway places. Our happy childhood days and teenage years of cooperative effort in caring for our bees, packing and selling honey, and playing real beekeeper were ended. They had been good years. For our whole family, beekeeping had been fun!

2

A Wonderland For Bees

IN MARCH 1950, my father and I took a trip to Santa Cruz, California. We instantly decided that this was the place for us. We moved to Santa Cruz within a few weeks, "lock, stock, and barrel," as my father exclaimed with satisfaction as we loaded the last of our equipment on a truck for the final trip from Davis, California to Santa Cruz. Wild flowers were blooming everywhere. Many of the eucalyptus trees were in full bloom as well as other trees and shrubs unknown to us. Honeybees were humming about their business wherever we stopped to look.

We could see at a glance that this appeared to be marvelous bee country for the production of wild flower honey. In support of our preliminary appraisal was the great diversity of flora in bloom with evident indication of more to follow as the season advanced.

Our first task was to find the most favorable location for our bees. Even in a land like Santa Cruz County where bees and flowers abound, we needed to keep in mind the fact that some areas would be more productive than others. We wanted more than honey. Almost anywhere we located our bees we could obtain honey. What we wanted was the absolute ultimate in location for maximum production per hive. Quietly we searched for this ideal spot. We drove to many areas of the county at various times of the year, stopping often to observe the bees and flowers. We hid "catcher hives" on publicly-owned park lands so that we might conduct experiments.

We learned that statistically Santa Cruz County had a reputation for being a poor, even less than average, honey producing area. Several times our local newspaper had published reports that the average production per hive was only twenty-five pounds of honey per year. We believed these reports were true. However, we were equally convinced from our preliminary observations that it was not a lack of nectar producing flora in this area that was responsible for the low honey yield, but rather a lack of ability on the part of local beekeepers to encourage their bees to gather the available nectar. We decided to continue and expand our experiments.

The day came when we had a chance to buy a little ranch in a lovely quiet setting. We bought it the same day we found it. This proved to be a wise decision for a few days later another man, who had seen the property just days before, came with money in hand to purchase the land for himself. He was too late—we already owned it. This incident pointedly illustrates one of the primary rules of beekeeping; namely, in all of our thinking and actions we must keep one step ahead of our bees. When we know that a hive needs a super or other attention, we must do what is needful without delay. Today's need, if delayed until next week or next month, seriously jeopardizes our chances of obtaining maximum honey production.

The ranch we purchased was in an ideal location, more so than we had dreamed was possible. We were directly under the flyway of wild bees from two directions. Soon we saw wild swarms pass over us flying from north to south, and from east to west. We put up some empty hives. Within weeks they were all occupied by swarms of bees. The thought my father had expressed almost twenty years earlier now really impressed itself upon us, "Why not try for a new world's record?" Why not indeed!

Many people who have bought honey from us through the years have expressed a desire to try beekeeping for themselves. Many have asked me, "Do you think that I could learn to keep bees?"

My answer always, "I would think so. Why not give it a try?"

Their reply, "Why not write a little book and tell us some of your experiences, how to begin, and some secrets of how *you* handle bees?"

I had no time to comply with their request until I sold my little ranch in Santa Cruz in the fall of 1971 to some dear young people who also bought five hives of my beloved bees. They had never kept bees before but sincerely desired to learn how to care for them and get delicious honey such as they had bought from us for years.

For these young people and for others who desire to keep bees I shall endeavor in the following pages to disclose some of the secrets of beekeeping that my father and I have learned from personal experience. It is my desire to help others to compete with my father and me in the race to set a new world's record.

3

Initial Preparation & Location Of A Hive

In the old days, all new hive equipment purchased from an apiary supplier came in knocked-down (disassembled) form, precut and ready to assemble. All the boards and nails to build the hive; its bottom, its cover, and frames were included in the package, but the buyer had to nail them together. Hives can still be bought in package form but now they may also be purchased completely assembled although at a higher price. If there is a need to economize and a person has natural woodworking ability and tools, the best thing to do is to buy one complete knocked-down hive and one complete knocked-down super and build his own additional hives and equipment as needed. My father and I enjoy making our own equipment, sometimes with various experimental deviations from the norm. Basically we use the standard Langstroth hive, a wooden compartment sixteen and one-quarter inches wide, twenty inches long, and nine and five-eighths inches deep outside measure. Adding a top and bottom makes the overall height about twelve inches. It is equipped with ten moveable and easily removeable frames. These frames do not take up the entire width of the hive body, some space being left on one side so that the frames may be more easily removed when occupied by a swarm of bees. Farm catalogues, such as those available free from Ward's or Sears Roebuck, as well as reference books obtainable at most libraries give all of the dimensions for building both the hives and various sizes of supers as well as much additional information useful to a beginning beekeeper.

My father and I make a departure from the usual factory assembled hive bottom in that we run all bottom flooring boards lengthwise of the hive instead of crosswise. This necessitates nailing a three-quarter by three-quarter inch wooden ripping all around three sides of the bottom to fit the hive body when it is placed upon the hive bottom. The fourth side, at the front of the hive, is left open as a doorway for the bees to use as an entrance. We find that running the bottom boards lengthwise makes it noticeably easier for the bees to keep their hive clean and dry—both vitally important factors in the production of a maximum amount of honey. We always raise the back end of the hive an inch higher than the front so that if rainwater blows into the entrance during a storm it can drain out by gravity.

Also, we never cut the usual four handholds into the wooden sides of the hive. Rather, we securely nail a three-quarter inch thick by two inch wide, by sixteen and one-quarter inch long, wooden cleat flush at the top at each end of the hive body or super to serve as handholds. Honey is heavy and such full-length handholds greatly speed up removing full supers and adding empty ones as the season advances. We nail such cleats to the bottom ends of each hive and super as a convenience so that if we want to set an empty hive on end it will stand upright.

We use medium depth extracting supers for best results in Santa Cruz but in many areas of the world full depth supers are better. Full depth supers are used to advantage where the weather is very warm or hot and the hives are located in a crop area such as a bean field, alfalfa, or cotton field. In these locations large amounts of nectar are gathered each day and the bees need much storage space. With full depth supers the beekeeper does not need to service his hives nearly as often as with smaller supers. Also, all hive bodies and supers, in whole or in part, are interchangeable at a moment's notice for hiving new swarms or adding more room to a full hive.

Factory-built hives generally come with what are called dovetailed joints. However, I have found half-lap joints are more easily cut and are equally satisfactory. Bees in plain box-type hives make as much honey as in any other, though this construction may not prove to be as durable as other types. Most of the straight-grained wood for the frames may usually be obtained free of charge if we make inquiry at

Exploded View of Hive

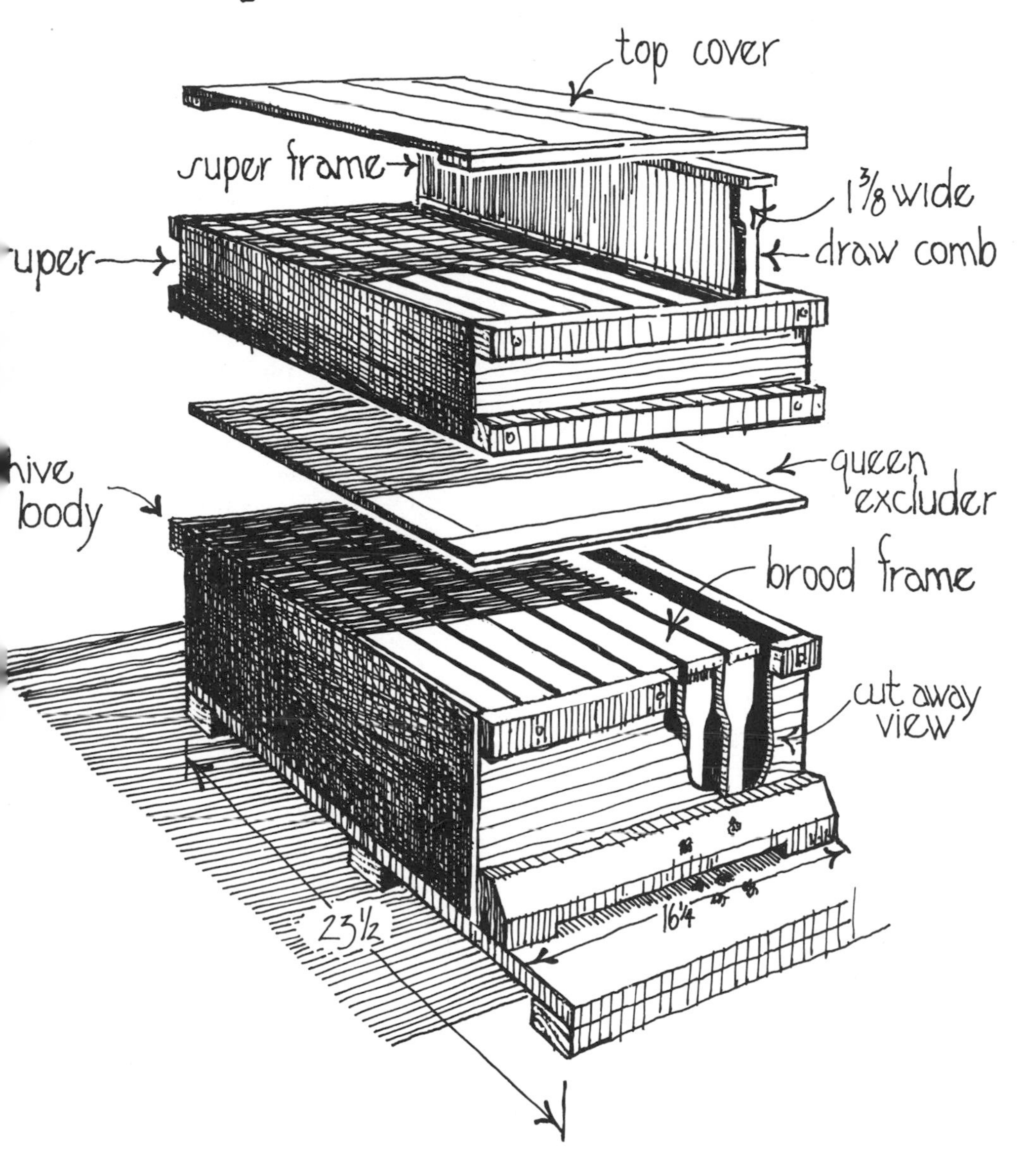

local manufacturing plants or furniture sales outlets. For instance, most large stores are glad to give wooden packing and crating boards to whoever will haul them away. Furniture factories have unfinished waste edgings and short boards that they give away as firewood. These kiln-dried waste ends are ideal for making beehive frames. We have a factory here in Santa Cruz from which we have obtained valuable free scrap lumber over a period of years.

The standard hive body has ten frames. Each of the frames must be supplied with what is called wire-reinforced brood foundation, or the recently available plastic-reinforced brood foundation. Both types serve the same purpose. These thin sheets of reinforced wax help the bees to quickly draw out some comb into which they may store nectar or pollen. These sheets also almost compel the bees to build nice straight honeycombs making it easier for the bee inspector to make his inspection. Foundation comb normally consists of sheets of pure beeswax embossed on both sides with hexagonal cell bases of the size bees built for worker brood. Our latest shipment, received only a few weeks ago from Diamond International Corporation, had an added attraction: the wax rims of the cell bases were raised almost one-sixteenth inch. Our bees loved this feature and immediately began to draw out these combs to their full length. The underside of the top bar of each frame is rabbeted out so that a wooden pressure plate (cleat) may be pressed against the wired brood foundation to hold it firmly in place. We use five five-eighths inch nails evenly spaced to secure the pressure plate to the top rail of the frame. Not all hardware stores carry five-eighths inch nails but the larger ones do, especially those that stock industrial supplies. These nails are packed in small cardboard boxes. It is important to use the correct length of nail to secure the pressure plates to the frame backs. If the nails are too long and protrude through the top of the back they must be filed off level with the wood or one will later strike them with the uncapping knife and ruin its razor-sharp edge while cutting off wax burr-comb.

Caution: When using factory prepared vertically-wired brood foundation (we use no other) be sure the *hook* ends of the wires imbedded in the wax are *up* under the pressure plate.

Full depth supers require much the same type of starter foundation as hive bodies. Medium depth supers such as we use may be supplied with either wired or unwired sheets of starter. We always use the unwired pure wax super starter. The reason for this is that we want to sell part of our honey production each year as cut-out comb honey. Super starter comes in various widths and we order the width to fit our frames. Wired or unwired foundation or starter is always of the same length. Much experimental work has been done to perfect the manufacture of completely drawn brood foundation so that the bees need not expend effort or time in drawing out the combs prior to the queen laying eggs or the worker bees storing nectar and pollen. Such plastic frames with partly drawn honeycombs are now a reality and may be purchased from some supply stores if one can afford the price. No wood or nails are used in making the frames. All is made of plastic coated with a very thin layer of pure beeswax. Some of my friends have used these frames for the first time this season and report excellent success with them. Only time will tell if they will prove to be greater honey producers than our present equipment.

We use frames with split bottom rails. This means that the bottom rail of each frame is made in two parts with a narrow space between. Split bottom rails are particularly useful when we use unwired pure wax started for our supers. We order wax starter of sufficient width so that we may bend the top one-quarter inch edge of the starter sheet over at right angles so the sheet is held firmly in place by the pressure plate, at the same time allowing the lower edge of the sheet to extend down between the two parts of the bottom rail to hold that edge straight.

Actually, of late years we have constructed our medium depth supers one-quarter inch shallower than the standard supers of this type listed in the farm catalogues. The reason for doing this is because we have sometimes had difficulty in obtaining wax starter sheets of sufficient width to allow us to bend the top one-quarter inch over at right angles. With our method of construction we can use either five and five-eighths inch or five and three-quarters inch width starter sheets, both readily available. If one uses only wired super foundation sheets for extraction, one has no difficulty. The bent ends of the vertical wires hold the sheets securely in place under the pressure plates.

Caution: If for any reason the starter sheets fall out of the frames the bees will become frustrated and discouraged and one's chances for a world's record are practically zero.

Stand for Hives

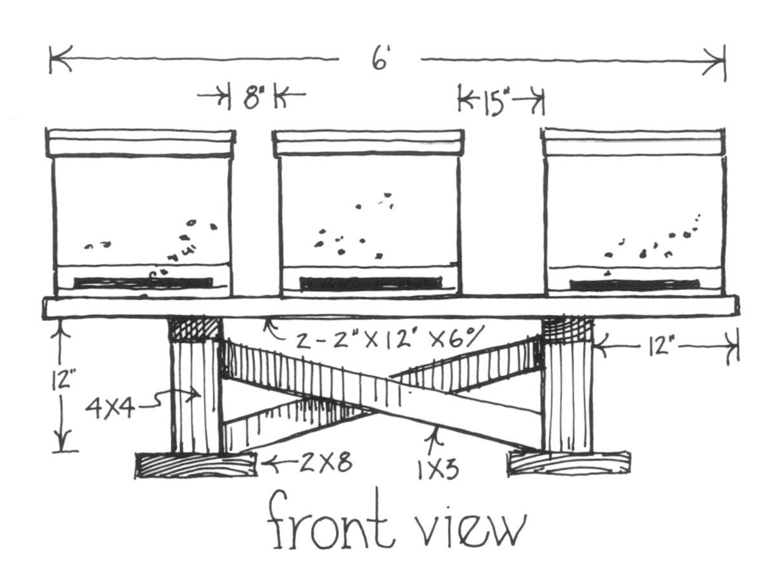

front view

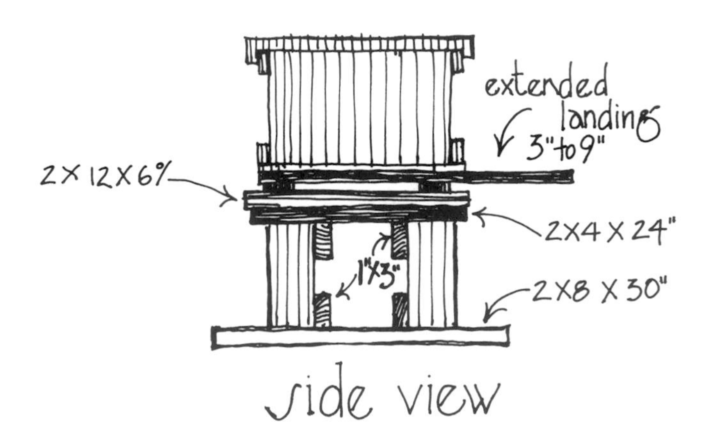

side view

Initial Preparation & Location Of A Hive

When a hive is all nailed together, we need to find a place to set it that will be most advantageous for the bees to work once they are hived and living in it. We have found that it is best not to place a hive on the ground, but rather on a stand at least sixteen inches high. The location of this stand is important. It should be placed where it is sheltered from the north wind and somewhat sheltered from wind from all other directions. The hive, when on the stand, should face east and a little south for best results. There should be at least twenty feet of open space in front of the hive to give the bees an open flyway when leaving or returning loaded. The front of the hive should get the full morning sun. This is not always possible but, for an ideal location, the above would be best.

A permanent stand consists of four four-by-four inch posts set one and one-half feet into the ground and extending upwards twelve or fourteen inches. They are placed in a rectangular position in pairs. Two are placed eighteen inches apart and the other two eighteen inches apart five feet away in a north and south direction. We place a seven foot four-by-four inch timber across the two most easterly posts and another seven foot four-by-four timber across the two westerly posts. The stand is complete when the hive is placed across the four-by-fours facing east. This stand will hold three hives evenly spaced. Large hives filled with honey are heavy and require strong timbers. We also build and use portable stands built of more lightweight materials.

In our latest experiment with permanent stands we have used two two-by-twelve inch planks seven feet long placed across the top of the four-by-four posts instead of the usual seven foot four-by-four timbers. The advantage of the planks is that they give us a wider solid base upon which to place our feet so that we can stand in a more comfortable position when taking off or adding a super. A possible disadvantage may be that ants and sow bugs have a larger hiding area under each hive.

A beehive stand saves much trouble. No ants can build their nests in security directly under the hives, eat the honey and bother the bees. Last spring a huge colony of ants built a nest in the ground near a grape vine not far from our Number 4 hive. I saw the ants climbing up the stand posts and going into the hive but I forgot about them for a few days. When I noticed them again the ants were entering in such great numbers that the frantic bees were trying to blow them out of the hive by fanning with their wings. They were even having some degree of success but at a great expenditure of labor on the part of the bees. I immediately ran for our bottle of ant poison, Kellogg's Ant Paste, and put a few drops on a leaf, then covered it by another leaf so that no bee could accidentally step in it. I did this in various locations near the hive entrance and also on the ground near the ants' nest. This was a most stubborn colony of ants and it was some days and only after repeated feedings of poison that they were subdued. Kellogg's Ant Paste may be purchased from the druggist at the prescription counter of most drug stores. We always have to ask for it, and sign for it, but it is the best ant poison we have found. The bees do not like Kellogg's, but we are always very careful anyway.

A stand also helps to keep the entrance of the hive free from grass and weeds growing up and interfering with the bees' flyway. This is very important, for if a bee hits a stout weed on the way in it may injure a wing and be crippled—and we need every last bee healthy and well to gather all the nectar possible during the honey flow—and that is the time when the weeds grow. Honey flow means that there is abundant nectar available for the bees to gather.

One day a man telephoned and asked me to come look at his bees. When I arrived he was busy for the moment but told me to go down to a certain roadway and along it I would find his two hives. I went down and soon found one hive but, look as I would, I could not locate the other. Soon he came down and I told him of my failure to find his second hive. He and the men with him all began to look for it. After a short search they found it where they had placed it in a shallow recess in the bank along the road. Tall milk weeds had overgrown the entrance so thickly that the hive was completely hidden. A few bees were making their way with great difficulty through the weeds to the entrance. After I had considered the situation I asked the men which one of them was going to clear away the weeds so that the bees could enter their hive more easily.

"None of us!" they exclaimed, as they all stepped back.

"All right," I said, "then all of you step back a little farther and I'll do it."

None of us had gloves or veils so I received instant compliance. It was not actually difficult or dangerous. I stepped to one side of the bees' flyway and began to talk to them in a low loving voice and at the same time began pulling weeds deliberately and steadily. In a few minutes all was clear. The owner and his men declared I must have some mysterious power over bees, for they had been stung until they were almost bluffed out of beekeeping. Bees are sensitive insects. If I had taken a hoe and hacked off the weeds in front of the bees they would have become frightened and stung me.

Most important of all, a good stand keeps the skunks from coming at night and eating bees. Skunks love to eat bees and they never get stung, at least not often enough to discourage them from eating bees. I sometimes have wished that they did. A few years ago we had ten experimental hives enclosed by a high board fence. We had some hives on high stands and three on a low stand ten inches above the ground. All hives were strong and doing nicely until one day we saw that the three low hives were noticeably weaker than the others. In three days' time the three low hives were very weak with few bees coming and going. We looked around and found a small hole under the fence such as a cat might use. Could it be skunks? To find out, we completely cleared all grass and weeds away from the entrances of these three low hives and spread a layer of dry sand all along in front. Next morning we had our answer—skunk footprints—big prints, middlesized prints, little prints and tiny ones. Old Grandma and Grandpa skunk had come and invited all their kith and kin to feast on our bees!

My father was angry. "We'll fix them," he said. "We'll put up an electric fence in front of the hive entrances and tickle their noses." We put up the fence and according to the footprints in the sand next morning we really had tickled their noses. The skunks had made a hasty exit but they had not stunk up the place as I had feared they might. There was no scent of any kind. We left the electric fence as it was with the power on. This was a grave mistake.

Two weeks went by and when the bees had made no progress in gaining in numbers, we began to search for the reason. After close observation we found it; the electric fence interfered with the economy of the hives. All hot electrical wires have an electrical magnetic field and the bees could not work efficiently in such a situation. Of course, we took the fence right down and raised the stand up out of reach of the skunks as we should have done in the first place. We still got some honey from those hives but not nearly as much as we should have. We learned by trial and error and sometimes our errors cost us money. Again, in our experience, one should never place his beehives directly under the wires of high-tension power lines. The bees will live, but they will not increase in numbers to the point where they can store excess honey for their keeper.

Let us talk about errors for a moment more. The great tendency among beginning beekeepers is to buy new hives, have some experienced beekeeper help them hive new swarms hanging from trees or bushes, and get started in the right way. They may possibly even get some honey from their hives the first season. So far so good. They feel elated! Now, why not accumulate a quantity of inexpensive old used hives and equipment and expand their beekeeping business? Soon people even bring them old hives and frames as gifts.

Caution: Never look for or accept these old hives or frames. Chances are too great that they are carriers of American Foul Brood, a deadly disease among bees, or other diseases or poisons. If you must accept such equipment to keep peace in the neighborhood, quickly cover it with a tarpaulin and bury it that night. Burning it, even at night due to the light, tends to attract your own healthy bees to the scent released by the heat and they may alight upon some contaminated part before you can get it all burned up. Then your bees are exposed to whatever caused the former occupants to die out of the old hives. Need I say more?

There are beekeepers in all parts of our country including our large cities. I have talked to people who live in high-rise apartment buildings in New York City who keep bees in screened-off areas right in their apartments. They leave a window open a crack and make an enclosed runway from the opening under the window to the hive entrance. Bees will crawl four to six feet through a two-inch tube to reach their hive. It is impossible for them to produce more than a nominal amount of honey in this situation, because it takes excessive effort on their part to keep air circulating throughout the hive to evaporate the nectar. With careful, intelligent handling, however, they may still make one or two supers of excess honey for their keeper.

When we move to a new location we make no effort to find out whether or not we can keep bees. There are so many federal laws, state laws, county and city laws and ordinances (current, obsolete, obscure, and otherwise) that it is almost impossible to determine the status of beekeeping in any given locality. What is important is to make the acquaintance of the local government bee inspector. He is our friend. He wants more people to keep bees. He knows what laws must be adhered to. Our local bee inspector sometimes asks me for the names of people I know who have bees. I always tell him because he *must* inspect, for the good of all beekeepers, all the hives in his area. An experienced bee inspector who loves honeybees and can handle them efficiently without fear is a good man to know as he has a wealth of practical information and know-how.

In cities or highly populated suburban areas, we must sometimes keep our bees camouflaged so that we may live in peace with our neighbors. As far as beekeeping is concerned, what our neighbor does not know will not hurt him. We place our hives (two or three) in an out of the way corner of the yard surrounded by a high hedge. Or we build a six or seven foot high decorative fence out of anything that is in harmony with the landscape. There is no entrance to this area where our bees are located except through a toolhouse or storage shed forming one side of the bee enclosure, or through a secret panel that opens into the bee space.

On one occasion we had a neighbor who took a violent dislike to me. The only way the old fellow could get at me was to report to the authorities that I had a considerable number of beehives. Soon I received a letter forbidding me to keep bees and further ordering that I get rid of all those I had, immediately. I wrote the city officials a nice letter reminding them that bees are quiet, inoffensive, useful little insects making no noise or muss as cats and dogs, horses or donkeys. Moreover, I kept my bees as a hobby. I also begged their permission to continue keeping a few hives. I heard no further objections from the city fathers.

However, we took warning. My father and I immediately built a thirty foot square enclosure with a small building as part of one side. We built this fence out of old redwood shakes and more than six feet high. The fence had four secret panel openings, one on each side. In this instance, we made no opening from the small building into the bee enclosure because I knew the old man would snoop through the unlocked building to try to find the door to the enclosure. We planted grape vines around the fence, soon almost hiding it from view. Inside we placed some of our best hives. The high fence caused the bees to fly away and return at a steep angle so that our shortsighted crosspatch friend was unable to see them. Peace was restored in the neighborhood. The local children knew our secret, of course, but none of them told the old man. Children love a secret and before long some of them had bees of their own.

Swarming bees can be handled quietly also. Before they swarm out of the hive, they fill up on honey and carry as much more in their honey sacs as possible. This makes them very gentle and they seldom sting anyone at this time. If they have already located a new home, in a matter of minutes the entire swarm flies away and our neighbors are rarely aware of what has taken place. If the bees have not found a place to go, they will soon settle and cluster, generally on our own property, provided we have made some provision in advance for this contingency. Bees like to cluster on young trees—apple, plum, apricot, oak—and on the wooden posts of trellised berry vines. We plant such trees and vines in various places on our lot or acreage. I have seen bees swarm out and cluster on an apricot tree only three feet tall. The poor little tree bent completely to the ground under the weight of bees. On the other hand, I have never seen a cluster of bees on a walnut, pear, willow, or bay tree.

Low hanging swarms are the ones that cause young mothers to fear for the safety of their children if a queen bee becomes tired during her flight to a new home and alights with her swarm to rest on a bush or tree in a back yard where children are at play. We always have a spare hive equipped and ready to catch a new swarm at a moment's notice in case of such an emergency. If we do not want the swarm ourselves, we leave it clustered where it is and telephone other beekeepers we know and tell them to come hive the swarm. Someone always wants bees and this word gets around among members of the local beekeeping fraternity. Backyard beekeepers are friendly folk who will do their best to hive unwanted swarms in as rapid and humane a manner as possible both for the sake of the homeowner and our friends the bees.

4

Inner Life Of The Hive

THE QUEEN BEE is the mother of all the bees in the hive whether there are 15,000 during the period of semihybernation during the wintertime or upwards of 125,000 keeping house in a large hive in summer. The entire economy of the hive depends upon her. Every one of the thousands of bees knows this fact. Therefore certain worker bees, sometimes called the queen's escort, are delegated to care for their queen. They bring her food when she is hungry, water when she is thirsty, bathe and cleanse her, caress and comfort her, warm her with the heat of their own bodies as they press closely around her if the hive temperature suddenly drops, or again fan and cool her if she becomes uncomfortably warm. In times of danger all of the bees attempt to provide security for their queen even though it may mean certain death for them.

In the daily routine of the hive the primary task of the queen is to lay eggs, sometimes as many as 3,000 a day during the early spring or summer buildup of the colony. It is the queen's attendants, caring for her every need, who make it possible for her to lay such a vast number of eggs. Before she can lay an egg the queen must lower and squeeze her body, tail first, into a newly cleansed empty cell so as to deposit the egg at the bottom. Then she must draw herself out of the cell and quickly repeat the act in the next available empty cell. This strenuous activity is very tiring for the queen so the escort bees continually surround and refresh her as she labors to lay eggs. Often we can determine the location of the queen on the surface of a brood comb even when she is almost hidden in a cell while laying an egg by noting the ring of escort and nurse bees facing the entrance to the cell occupied by the queen. My father and I look for what we call the "daisy pattern." The queen's attendants are the petals and the queen herself the center of the flower.

After the queen has laid many eggs and the hive has become overcrowded the worker bees build one or more (usually three or four but on rare occasions as many as eighteen) extra large specially constructed cells that hang downward with the opening at the bottom of the cell. In these the queen lays an egg which if properly and tenderly cared for will produce a new queen. Soon after the egg is laid the nurse bees begin to supply the cell with a specially prepared potent food component called "royal jelly." In three days the egg hatches and the young larvae begins to feed on the royal jelly. It grows rapidly, at the same time developing those organs and characteristics that in sixteen days will cause it to emerge from its cell as a beautiful, slim, virgin queen. On the next warm sunshiny day she will usually find her way to the entrance of the hive, take wing and fly away up into the blue of heaven. At a height of forty or fifty feet a number of drones will find her, they will mate, and she will return to the hive a vigorous fertilized young queen ready to take over the duties of the colony. Sometimes just before the young queen emerges from her cell and flies away to be mated, the old queen completes her preparation to ease the overcrowded condition of the hive by calling approximately one-half of the adult population to follow her out of the entrance into the shimmering summer air to form a swarm to go in search of a new home. Often this great buzzing, swirling of thousands of bees soon find that their queen has flown away but a short distance from the parent hive and alighted upon the branch of a tree, fence post, or other object. Then they also alight over and around their queen in a great cluster above and below, where they may remain for a few minutes or a longer period of time depending on how soon they can find a new home or a beekeeper comes along and hives them.

Caution: After hiving, remember that bees need air. Never at any time plug the entrance tightly with grass, blocks of wood, or other materials, but confine them with a full width screen closure.

Queen Cell
Partially Completed

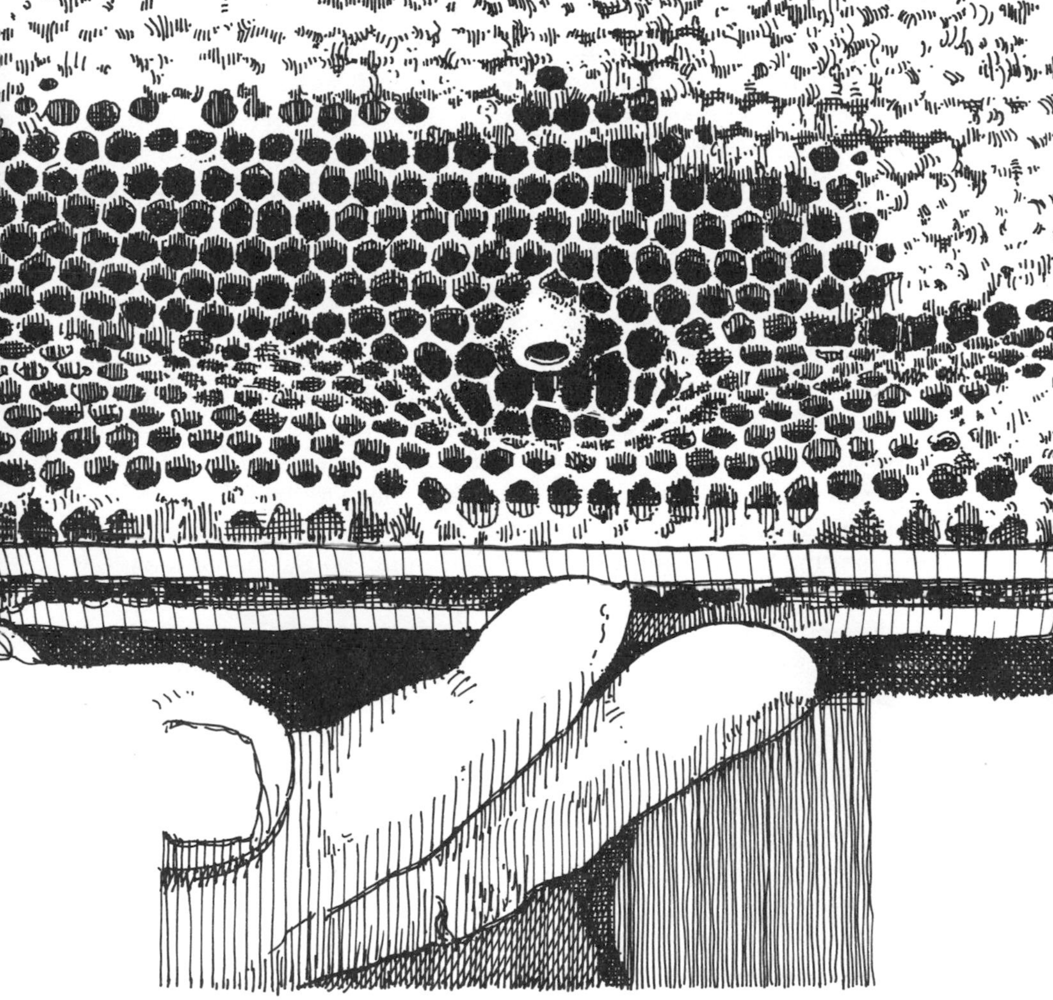

Every year I hear most disheartening accounts of people who have tried to pen bees in a hive prior to moving them. Surely, they think, blocking the entrance for an hour will not hurt bees. But it does! They become frantic for lack of air, the hive temperature rises from 95 degrees to 160 degrees—the scalding point of water—or even higher as the bees in a large hive succumb for lack of oxygen. The high temperature melts the honeycombs killing all of the brood, and the honey runs out of any crack in the hive onto the ground. Of course the whole comb interior is also a total loss. The same tragic result occurs when someone places a beehive in a large plastic bag, or covers it with a plastic tarpaulin to confine and protect bees when the owner has been notified that an orchard where the bees are located is going to be sprayed. Move your bees out of the orchard when notified that spraying is to be done. Better yet, do not place bees in such a hazardous position, but locate them either upwind from the orchard or off to one side of it, or out in uncultivated areas as we do.

After the swarm has been successfully hived, the cover replaced and securely tied, and the new bees taken home and put on their stand, the bees go right to work to make the hive a permanent home. The frames in the brood chamber are manufactured so as to be self-spacing and should be pressed tightly toward the left side of the hive. This enables the bees to work at maximum efficiency and checks their tendency to build spur comb between the frames. As a result the great bulk of the bees usually move to the left side of the interior and take possession of one, two, or three frames, depending upon the size of the swarm and begin to draw out the foundation comb. Since it takes close to 100 degree heat to draw honeycomb the bees often move to the warmest side of the hive. Thus if the hive entrance faces east the bees begin to draw the frames on the left hand side as one faces the entrance, or the side of the hive facing south where the sun can warm it most of the day. A strong colony begins to draw comb within a few hours and from then on one can hear the sounds of construction both day and night during the winter, spring, or summer, depending upon the geographical location of the hive.

However, not all of the bees are employed as homebuilders. Some are field bees and they almost immediately come out of their new home and fly around and around the entrance to locate themselves in relation to their new surroundings so that when they fly away in search of nectar or pollen they can find their way home again. The older experienced field bees accomplish this purpose in fifteen minutes and fly away to work. The younger bees take longer to assure themselves of their location. That first afternoon we will see a few loads of pollen being brought to the hive but not many as the bees have not yet drawn sufficient honeycomb cells to store much pollen or nectar. The first cells, even though only partially completed, serve as depositories for the honey the worker bees had brought with them from the parent hive as they always fill their honey sacs before swarming so as to insure a three day's supply of food in case of need. They are wise enough to know that their scout bees may not be able in one afternoon to find a permanent new home, so they take a maximum amount of honey with them.

On the second and third days many more of the homebuilder bees are sent out as field bees to gather the necessary stores, so we see many of them flying around to familiarize themselves with their new habitation. As the days pass we see many, often huge, loads of pollen and nectar being brought to the hive. Since sufficient storage space is now available the field bees work from dawn to dusk, weather and temperature permitting, to store quantities of nectar and pollen in the prepared cells ready for the hive and nurse bees to use as needed. Also on the second or third day the queen bee begins to lay eggs, often in only partially completed cells. After she has passed on to the next cell to lay another egg the hive bees continue to build the cell so that in a few days other bees can store a mixture of nectar and pollen in the cell after the egg hatches so that the young larvae will have something to eat. The bees know they must exert every effort to encourage their queen to lay eggs as soon as possible so there will be young worker bees emerging in twenty-four or twenty-five days from the date of hiving to replace the field bees who literally work themselves to death in six to eight weeks at this time of the year.

Field bees arise early in the morning and a few of them fly out of the entrance to test the state of the weather. If the temperature is forty-five degrees or less they quickly return to the hive. But as soon as the morning sun warms the entrance and surrounding air to fifty-five degrees or more, many field bees take wing and continue their endless search for the necessities of life. During our winter buildup of the hive here in the coastal regions of California we sometimes have days of sunshine interspersed with periods when large clouds suddenly obscure the sun. At such times, as the cloud's shadow passes over the front of the hive, we see an immediate lessening in the number of bees leaving and a noticeable increase in the number of bees returning whether loaded or not. I have marveled at the bees' ability to forecast the weather on short notice. If the big cloud overhead is going to drop some rain the bees stay in or near their hive, but if it is just passing they continue their work flights.

As the days pass the hive bees continue to draw more honeycomb, working in turn on each adjoining frame. A fine strong hive can draw all the starter sheets in all ten frames in a period of three weeks, though often it may take them four weeks in our area. This is a truly remarkable accomplishment considering that during this time no young bees can possibly be added to the work force because it takes twenty-one days for an egg to hatch and emerge from the cell as a young adult. Meanwhile the total work force has constantly been diminishing due to the death of the aged field bees, but the remaining bees work on, sometimes it seems almost in desperation, until the first of the new brood emerges from the cells. When this happy day comes we see an unmistakable air of rejoicing all around the entrance. And as beekeepers we rejoice with our bees. Now our hive is on a secure footing and we must think about placing a queen excluder and adding a super for additional storage space in a matter of days. This is a crucial period in the life of any hive. If thousands of brood emerge every day the interior soon becomes overcrowded, the workers build queen cells, and the bees swarm.

My father and I use what we term the "hollow tree" theory in deciding when we should add a super to a beehive, whether it be in late winter, spring, or summer. As a rule, in our part of the world, wild bees live in hollow trees. When wild flowers begin to bloom bees build up rapidly in numbers and a small hollow in a tree soon becomes completely filled with pollen, honey, and bees, causing them to prepare to swarm. Now if the hollow in the tree is large, the bees continue to build comb and increase in numbers for a longer period of time before they think of swarming. They like to have their home filled to capacity with the good things of life before they leave it, forced out for lack of space.

We think of each of our hives as being a hollow tree. When we see our bees crowding the entrance in the early part of the season, we first remove the restricting cleat so as to enlarge the doorway. If this maneuver seems to do the trick we wait yet a few days until the entrance is again being crowded. Then we add a super, the equivalent of making the hollow in the tree larger to accommodate the increasing number of bees. Again and again we watch and count the number of bees coming in for a landing at the entrance and each time, when necessary, we add another super to enlarge the hollow in our theoretical tree even more. I am sure our bees appreciate what we are doing for them because after dark we can press an ear against the empty super we placed on the hive earlier in the day, and hear a joyous whirring of wings as the bees prepare the combs for the storage of honey next day. Our bees are pleased at not being forced to move and we rejoice at the abundant golden harvest they tuck away in the supers for us.

Earlier in this chapter I mentioned a queen excluder. This is a device used to confine the queen bee (also the drones) to the lower part, brood chamber, of a beehive. The excluder somewhat resembles a wide-mesh screen fitted into a wooden frame with the same outside and inside dimensions as the top edge of the brood chamber. It is a precision piece of equipment made of stainless steel wires placed .163 inches apart with four reinforcing cross wires evenly spaced. The opening between the wires of the excluder are a tiny fraction of an inch too narrow to allow the queen to slip through. If left to her own inclinations the queen bee will usually circulate throughout the interior of the hive, laying eggs in various places from top to bottom, sometimes making the topmost supers her principal brood chamber. Such a roving queen makes it more difficult for a beekeeper to service his hives. He must always use more care when adding or removing a super to be sure that he does not injure the queen. If he accidentally brushes her off a comb and she falls as little as one inch directly onto the wooden back of a frame, she may be injured permanently and her ability to lay eggs drastically reduced.

Every new queen excluder should be examined to see that it has been properly assembled at the factory. The wooden frame around the stainless steel should be neatly joined at all four corners. If any joints are uneven the staples must be pulled from two corners and the loosened side board turned end for end. Then it will fit properly into its precut slots and may be restapled. Also, we often find four small holes, one at each corner of the excluder. These must be plugged with wood, beeswax, or putty. If such holes are not closed they act as four small chimneys to let the warm air escape from the hive. In cool climates bees can make little, if any, excess honey when burdened with this handicap.

Queen excluders are a real boon to the beekeeper and do not seem to be a serious hindrance to most bees. One reason for the bees' ability to work efficiently through an excluder is the fact that the heavily laden homecoming bees entering the hive do not try to pass through the excluder to carry their loads to the topmost super. Rather, they deposit their loads of nectar or pollen in the most readily available cells in the brood chamber below the excluder and the hive bees, young slim workers for the most part, pick up and carry the loads through the excluder to higher parts of the hive. Honeybees have refined the entire domestic operation of the beehive to a fine art.

Once in a while, as with human beings, we find bees that are larger than average. Such bees cannot work through a queen excluder. Three years ago we unknowingly hived such a swarm. The first season they produced no excess honey, but built up into a fine colony. Last season they should have produced excess honey for us but did not. Early in the season we observed what seemed to us to be peculiar activity on the part of the bees. We were deeply puzzled. They were not lazy for they built up rapidly in early spring, became overcrowded and swarmed. We hived this large swarm. In less than a month the parent hive had built up again to the point of swarming a second time. Though we had given them a queen excluder and two supers with drawn combs, they made no effort to utilize this additional space. We raised the excluder and slipped one of the supers below it, thus enlarging the brood chamber. The bees immediately began working in that lower super but still not in the upper one above the excluder. We began to have an inkling of the problem. We exchanged the old excluder for a new one thinking the manufacturer might have made an error in spacing the wires. But there was no change for the better. The season passed without profit to us, at least in the form of honey.

This spring we placed a queen excluder on both the parent hive and its swarm of last year. Neither one could work through an excluder. After two weeks we removed both excluders and the bees went right to work in all the supers we added. We requeened the secondary hive early in the season so that we will be able to use an excluder on that hive next year, but have kept the parent hive as it is for an experiment. This year each of the two hives will produce at least one hundred fifty pounds of honey—by no means a world's record but still a goodly surplus for us.

5

Ways To Aquire Bees

THE MOST LOGICAL WAY to begin beekeeping is to buy a few good strong hives, or "stands" as they are often called, from an established beekeeper. However, at the present time few beekeepers want to sell bees; rather, they want to buy more for themselves. The reason is that they have more and more call for honeybees for pollination purposes. Another way to start beekeeping is to buy the hives and equipment from the aforementioned Montgomery Ward or Sears Roebuck farm catalogues; or from Diamond International Corporation Apiary Division, Chico, California. This company also has a free bee supply catalogue available upon request. One can assemble the equipment, set it all up, and order package bees. Ordering the equipment is satisfactory but ordering package bees is sometimes a disappointment for the reason that the suppliers of bees may send the bees after our coastal region honey flow has almost ceased. Then one's package of bees may or may not be able to build up strong enough to gather sufficient honey to supply themselves for their own winter's needs. If they do not, they must be fed. I will come to this later on.

The best way to get started, we have found, is to assemble the equipment, get the hives ready and set up for the bees to occupy by April first. Then telephone, or better yet, personally visit some of the fire departments nearby and tell them we want to hive swarms of bees. Householders continually ask firemen and policemen to help rid their properties of wild swarms. These public officials are glad to know of someone who wants bees, but one should not bother them until the swarming season in their particular area has actually begun.

When apple trees come out in full bloom, bees begin to swarm. All preparations for hiving swarms should be completed prior to that date. If one already owns one or more hives and his goal is maximum honey production rather than to increase his number of hives, he must place supers on the strongest hives at least three or four weeks prior to the apple blooming season.

One year my father and I wanted to acquire fifteen new swarms. We contacted the fire department, police department, and animal shelter people in our area. Soon after April first, we began getting telephone calls and by May eighteen we had all fifteen new hives filled with new swarms. These swarms may be obtained without cost to the beekeeper and are particularly desirable because we have found them to be free of disease.

Where do bees come from? Bees live for years in large hollow trees, or in walls of houses or old outbuildings. One of their favorite nesting places is in a permanently abandoned chimney. Bees will also sometimes move into a chimney after wintertime use when the fireplace is no longer needed in spring. An elderly lady of my acquaintance always telephones me about March fifteen to see if I have seen any new swarms. If I have, she says that she must then lose no time in having an iron cover put over her chimney top or the bees will surely move in. They seem to like her chimney even more than others.

There is no easy way to get them out and hived that I know of, nor is there an easy way to get bees out of a wall in a house and into a hive. The risk is not so much in getting the bees out and hived as it is in leaving the roof, wall, or other part of the homeowner's premises in satisfactory condition. Beginning beekeepers will do well to turn down such offers of bees.

For the advanced beekeeper there are two ways to remove bees from such locations. As an example let us consider removing the bees from their permanent home in the wall of a house. First, if we live in an area of the world where in late spring and summer warm days are consistently followed by warm nights, where the daytime temperature is always one hundred degrees or higher and the nighttime temperature not lower than ninety degrees, we use this method. We go to the house where the bees are in the wall, find the place where they enter and, with our veils on, build a secure platform just below the hole or crack through which the bees enter the wall. We remove from a strong hive we already own two fully drawn brood combs filled with honey, pollen, young brood, day old larvae, and eggs. These, with their adhering bees, we take from the parent hive and place them in a newly prepared hive, being absolutely sure that the queen of the parent colony is not among the nurse bees clinging to and covering the brood and eggs.

We take this new hive with its nucleus of bees and brood and set it on the platform already prepared, with its entrance so close to the wall opening that the extended landing board extends under and beyond the wall entrance. Then we fashion a twelve-by-eighteen inch piece of window screen into the shape of a cone with the large (at least four inch diameter) end tacked securely over the hole in the wall with the hole being as nearly as possible in the center of the cone. If the wall has an elongated hole or crack we partially plug it with soft clay or putty to reduce the bees' entry to a size that can be covered by the wire cone. The small end of the cone has an opening approximately one-half inch in diameter, five or six inches out from the surface of the wall. The field bees will readily find their way out of the small end of the cone and fly away to the fields in search of nectar and pollen. When they return they will land everywhere on the wall of the house near their usual opening and all over the entrance and landing board to the new hive with its combs of brood that we have just set up. These returning bees will try in every possible way to reenter their old home but the screen cone prevents this. The poor field bees feel lost for awhile and then, sometimes late in the afternoon, they begin to enter the new hive placed ready for their use. As a rule they do not fight with the bees covering the brood in the hive because the hive is at the moment queenless. Without loss of time the combined bees begin to rear a new queen from one of the eggs in the brood comb cells. Next day, more and more of the field bees leave the wall through the cone and more readily find their way into the new hive. We leave our new hive on its platform for one month. During that time all the brood in the wall of the house hatches out and becomes field bees, and almost the only bees left in the wall are the queen and her escort. At this point I plug the entrance in the wall for at least a week, then open it to see if the new hive bees will enter and rob out the honey stored in the wall. If there is still a honey flow the bees will not rob out their old wall hive and I must again plug the entrance for a longer period even though there is a risk that the wax moth eggs and larvae in the old combs may do damage to the wax and honey while we wait. But as soon as it is late enough in the season for the honey flow to have ceased I reopen the hole and the bees remove the honey from the wall in a few days, provided we have given the new hive a queen excluder and additional supers with drawn combs as needed. For bees to store the honey from the wall as fast as they are able to remove it we may need three extra supers and be prepared to extract these several times within a few days. Swarms living in houses may have little excess honey or many gallons. The beekeeper must be careful to observe, when he reopens the hole in the wall to let his hive bees rob out the old chamber, that it is his own bees that are doing the robbing and not some professional robbers from an unknown hive.

Caution: The above method must not be attempted when the weather is so cold that there is danger of the young larvae or eggs becoming chilled so that the bees cannot rear a new queen.

If the weather is too cold to remove frames of brood and eggs from a hive, we use our second method. We await our chance to hive a very small swarm of bees—preferably one with only three or four handfuls of bees, and a queen. These we hive into a standard hive with full sheets of foundation comb and take it home for six days until the queen has had time to lay some eggs in a small patch of newly drawn comb. We then take it to the house with the bees in the wall and place it on the same type of secure platform as in the above example, using the wire screen cone and all else as described above. The new queen, having but few workers, will usually not set a guard. Again after waiting around for some hours the returning worker bees will enter the newly placed hive and take up housekeeping with the new queen and her little band. Here again we must be ready quickly to add more room to the hive as it is needed to accommodate a large swarm. Such an adventure requires skill on the part of the beekeeper and an unusual amount of patience on the part of the homeowner. However, there are many kindhearted people who will work in harmony with a beekeeper in an effort to save the lives of our precious little friends.

Wild Swarm

Another way to get started with bees if one lives far out in the country is to look for wild swarms. First, look carefully at all the flowers or trees that are in bloom to see if there are any bees working on them. If there are no bees to be found at all, the chances are slim that a wild swarm can be caught. However, if bees are at work on blossoms here and there, this indicates that there is a hive or bee tree somewhere in the area. These bees will probably swarm at least once in the spring or summer so it pays to keep both eyes and ears open. Neighbors willingly watch for swarms flying overhead if we suggest sharing a little honey with them at a later date. If we or a neighbor see a swarm pass, we note carefully the bees' line of flight and try to get an exact fix on the direction by noting two landmarks over which they have passed. Bees always fly in an almost straight line and then may be followed and found. In years gone by, my father received telephone calls from neighbors who had seen a swarm pass as much as an hour earlier. He would follow their line of flight over hill and dale and would find them nine times out of ten, usually within half a mile. Often the bees were clustered where he could hive them immediately. Occasionally he would find where they had taken possession of a hollow tree. In this latter case he would let them alone until late summer, until they had gathered as much honey as possible. If he could get permission to do so, he then cut the bee tree. At times he got a stinging and a quantity of good honey. More often his only reward was a good stinging! Let me state here that bees definitely do not appreciate having their tree cut down and sent crashing to the ground even though done with the best of intentions—to remove them into a nice new hive.

A true "bee tree" is a tree of any species large enough to have a hollow somewhere in its trunk or a larger branch capable of housing a swarm of bees. The entrance to the hollow is usually a hole one to four inches in diameter residuum from a rotted out tree limb. But an old tree standing in an exposed location where it is often wracked by the wind may develop cracks in the trunk through which the bees find access to the hollow interior. If we decide to cut the tree in an effort to hive the bees and secure the honey we always note carefully the location of the bees' entrance. If it faces south we plan to fall the tree in either a southerly or northerly direction, or if the entrance faces east we try to fall the tree in an easterly or westerly direction. The reason for this is that in our experience bees always build their honeycombs with the ends of the combs pointing toward the entrance. Thus if we fall the tree either in the direction of the entrance or directly away from it, the honeycombs are placed in their strongest position to withstand the shock of the falling tree and we may find them in a condition good enough to cut out and fit into our hive frames. Of course we always try to fall the tree in such a way as to drop it with the least impact and destruction to the bees.

Old dead trees, called snags here in the West, are hardly worth cutting as without leaves and twigs they fall with such a crash that we find the bees, combs, honey, and rotten wood all dashed together in an inextricable mass. It is better to let such a snag stand and the next spring watch it so as to secure a fine swarm of bees from it as they emerge after they become overcrowded and often cluster on a low-hanging limb or shrub nearby.

Some years ago the owner of a large oak tree refused us permission to fall the tree but he let us climb to the hollow, about thirty feet up, and saw, hack, and hew a door in the tree to get at the bees and honey. We hived the bees, took the honey, and then nailed the door shut again. We got stung, of course, as we knew we would before we climbed up there. I have never had the opportunity to return to see if another swarm took possession at a later date.

On occasion, in our coastal regions of California, a large swarm of honeybees will cluster under a sheltering gable of the roof of a house or even under the trunk of a fallen tree. The scout bees having been unsuccessful in their search for a new home, the worker bees begin to draw honeycomb even in the exposed location where they have clustered and start construction of a permanent abode. A few days ago, June twenty-seventh to be exact, we received a telephone call from a man in the Aptos area, about ten miles from our home. He said he had been clearing encroaching bull thistles, berry vines and poison oak from a walking trail in the hills behind a large mobile home park when he came upon a fallen oak tree across the trail. The fallen trunk lay in a horizontal position about waist high across the walkway. He was about to begin sawing through the tree when he noticed a huge swarm of bees clustered under the trunk amid a tangle of small dead limbs and leaves. He headed for home and called for help. Several people who had wanted swarms of bees hastily declined his offer to give them the swarm. Then he called the local Agriculture department and they gave him my number and told him to ask for Ormond. He did. At first I was not too eager to go such a long distance for what I thought would surely prove to be a small secondary swarm since our swarming season was almost at an end, but I had on hand an empty hive belonging to a dear young woman who so much wanted to get started with bees and eagerly awaited my telephone call saying I had hived some bees for her. It occurred to me to ask the man to explain his situation. When he described his bees as a permanent "free-hanging hive," as my father and I call such swarms, I decided instantly to help him and said we would be right over. The prospect of such an unusual adventure does not occur many times in the life of a beekeeper and this was a golden opportunity to test our latest theories on hiving such swarms.

We hastily loaded our station wagon with the necessary pieces of equipment. These included a complete beehive with five of the ten frames supplied with wired sheets of foundation comb and the other five frames with no starter sheets, two six-quart pans, a bee brush, four two-foot squares of cardboard, a few foot-long pieces of narrow board, our smoker and burlap fuel, bee veils, household rubber gloves for me, a long thin butcher knife, and a quantity of relatively thin cotton string. When our caller, Mr. Georg Nilsen, saw us coming with such a strange assortment of gear he was truly puzzled. I lost no time in approaching the cluster to give them my love and see if they would reciprocate. They did without a moment's loss of time. It was indeed a large, beautifully symmetrical free-hanging hive shaped like a beach ball and, of course, brown all over because bees completely covered the entire outside surface.

Mr. Nilsen loaned me his handsaw and twig clippers. In a few minutes I had cut away all the intervening twigs and branches and the swarm was in full view. Mr. Nilsen had his camera with him and when I had finished exposing the bees he asked me to sit down beside them so that he could take a picture. I was glad to comply. He snapped the picture and many more as my father and I made preparations to hive the bees.

First we placed the hive on the ground on a sheet of cardboard two feet to the right of the swarm with the entrance facing southwest toward the receding sun. Then I positioned two more sheets of cardboard under the bees so that none of them as they fell from the swarm would be lost in the grass and leaves but rather could crawl on the cardboard toward the entrance of the hive. The thin boards I used to support the pieces of cardboard. I gave the swarm two very light puffs of smoke from our smoker so as to alert them to the possibility of fire but not disturb them unduly. Then I thoroughly smoked Mr. Nilsen, my father, and myself from head to toe to kill our body scent. The next step was to remove the hive cover and two empty frames. My father laid one empty frame flat on its side on the remaining sheet of cardboard and cut five lengths of cotton string, each one long enough to pass around the frame with plenty of length to spare so that in due time I could tie them in place. These strings he placed under the frame and waited for my next move. This was to hold one big pan close up to the bees and gently brush some of them from the cluster into the pan and then dump them out over the remaining frames in the hive. A second brushing of bees into the pan revealed a bright yellow honeycomb hanging from the oaken tree trunk. This I carefully cut off with my big knife, held the fragile comb in a horizontal position with the cut edge resting on top of the frames in the hive and gently brushed the adhering bees off onto the frames. I laid this first honeycomb aside for the time being as it was too small and fragile to fill a frame. Again I brushed bees into the pan and transferred them to the hive until the second comb was plainly visible. This was a large sturdily built comb with some honey, pollen, and much brood. Using the second previously removed and empty hive frame for a measure, I held it up against the hanging comb and marked out with the knife tip the exact size of the comb necessary to fill the frame my father had ready. Slowly and carefully so as not to injure any of the bees I sliced off the lower area of this comb, brushed the bees onto the hive, and discarded it into the second pan because this portion of the comb was too fragile to handle. Then I sliced off the main section of comb I wanted to save, caught it in both hands, again brushed off the adhering bees and lay the comb into the frame my father had prepared. Together we pressed the comb into place as it needed to be slightly oversize to assure a tight fit, tied the five strings around the frame and comb equidistant from each other and then raised the frame into a vertical position by raising one edge of the cardboard support. The comb held its shape nicely and I placed the newly filled frame into the hive. We repeated this procedure until all of the large strongly built combs were tied securely in place and safely lowered into the brood chamber. All of the odd shaped or extra pieces of comb I placed in the second pan to take home for future use.

After all the combs were cut off the tree trunk we removed the hive to a position directly under where the combs had been so that we could smoke and brush the remaining thousands of bees off the top of the various smaller limbs radiating from the tree, down onto the top of the hive. In a few minutes this was accomplished. Slowly we replaced the cover, and had our lovely bees—comb, brood, honey and all. Elapsed hiving time: two hours, with another half hour wait until the fanners giving the "We've found a home," call could persuade all the stray bees flying around to enter the hive. We locked the bees into the hive with a wide open screen closure cleat, tied a rope around each end of the hive, passed a stout stick through the rope loops and each taking an end we carried our beloved bees down the hill to our car.

As soon as we arrived at home and had the bees placed on their stand I removed the closure cleat, waited a short time and then slid a saucer sized piece of the unused honeycomb containing both honey and pollen into the entrance. We always do this so that the bees will have the extra food immediately available and also the wax comb to cut up and use in the urgent task of fastening the combs permanently in place to the backs of the frames and bottom rails so the cotton binding strings will no longer be needed. Five days after hiving, my father shared with me the good news that the bees had cut the first string that morning and dragged it to the entrance of the hive where he helped them to remove it. At noon the end of a second string became visible. As soon as the air became warm enough, about three o'clock in the afternoon, we quietly removed the cover from the hive and found the bees had completely finished their repair and secure work. So we cut all the remaining strings and removed them from the frames. Our bees rejoiced and so did we. It was a successful adventure, netting a fine hive of bees and no one had been stung or even threatened. Mr. Nilsen had taken his pictures, first from a short distance away, and then from right over my father's shoulder without wearing protective gear of any kind. The bees loved him too.

Advanced beekeepers, those persons who are well along in the knowledge and art of beekeeping, may find pleasure as well as profit from putting up what we call "catcher hives." These are hives we prepare when we want to add a few more hives to our apiary but instead of going after the bees, we want the bees to come to us. In the usual practice we fill all ten frames of our new hive body with either wire- or plastic-reinforced brood foundation, shake a swarm of bees over the frames and the bees go right to work to draw out the combs. However, it is rare indeed in our experience for a swarm of bees to take possession of this type of properly prepared hive of their own accord. If we deliberately place them in such a hive they will stay and draw nice straight combs but they prefer a different nesting place.

Catcher hives are prepared so as to attract bees to come and take possession all by themselves. The hive is of the usual size and shape in all details except one. Instead of putting in full sheets of brood foundation, we cut thin sheets of pure wax super starter into strips about one and one-half inches wide and attach a strip in each of the ten frames. If at all possible, we use frames that have already been used in a hive at least once so that they have the scent of bees upon them. This gives one a distinct advantage over a brand new hive.

We locate this hive in some sheltered, warm and quiet place, such as on the low edge of a roof at the rear of a garage or other outbuilding, with the entrance facing east. We build a little wooden stand to fit the pitch of the roof so that the bees will be on the level. What attracts bees to come to such a hive is the fact that bees really like to be able to *cluster* in their new home and the full sheets of foundation comb preclude such a possibility in a standard hive.

During the winter months we usually prepare three of these hives. A few days before bees begin to fly in the spring we place them on their stands. It will still be some time before any hive builds up strong enough to swarm. This is good, for it allows time for our new catcher hives to lose our human scent. As the spring days lengthen, we see a few bees inspecting the entrances to our new hives. Sometimes there are bees around the doorways of all three. This is a good sign. It means that within a mile or two there is a strong colony with queen cells almost ripe and a big swarm about to depart for a new home. The bees around our catcher hives are scouts looking for this new location.

Bees are smart and possessive. The same hive will often send out scouts to take temporary possession of each catcher hive we have ready. They will resist the presence of any other bees that may be out scouting around for another home. From the beekeeper's point of view this is not good, for we would like to have all three hives occupied as soon as possible by three new swarms. Nevertheless, we are always glad to catch what we can and that is generally one swarm at a time.

When the queen brings the swarm to take full possession of one of the catcher hives, the other scouts join her in their new home. Just who makes the final decision as to which of the three hives the swarm will occupy, I have not been able to determine. It is a marvelous sight to see a huge swarm of bees taking possession of a hive all of its own accord. First we hear a low roar—the buzzing of up to 40,000 bees—and the air all around and above us filled with bees flying around and around. It seems as though there could not possibly be any rhyme or reason to account for their actions, but there is! In a few minutes they begin to gather in greater numbers around a certain hive, then begin to enter. What a sight to behold—thousands and thousands of bees all alighting and marching into the hive in an orderly manner. Many bees, heavily laden with either honey or pollen and weary from their long journey, alight to rest upon anything close-by while awaiting their turn to enter their new home. The ground, the hive, and all objects near at hand seem to be covered with bees. We will be covered with bees too, if we stand within twenty feet of the hive they are entering.

One spring the largest swarm of bees my father and I have ever seen suddenly swooped over a high hedge on the north side of our property and stopped in the area of our three catcher hives. It looked for sure as if they would go into a certain hive, so Father and I moved up closer to watch. Suddenly they veered to the left and began entering a hive just eight feet from where I was standing and only twenty-five feet from my father. In a moment, before I could move two steps, I was covered with bees—literally hundreds of them. My father spoke in a low voice, "Ormond, you are just covered with bees from head to foot. Stand perfectly still or they will crawl up your pants legs or down your neck!"

It was excellent advice. I froze and tried to make myself as much like a marble statue as I possibly could. Out of the corner of one eye I looked to see how my father was faring. He, too, was covered with bees! However, in five minutes there were noticeably fewer bees outside, and in ten minutes there was not a bee left on either of us. As we continued to watch, we saw the orderly march of bees into the hive suddenly begin to slacken and then cease altogether. My father cried out, "Get another hive body quickly. There are so many bees the catcher hive is full. We must give them more room before they start coming out again to look for a larger home."

I ran for our storage shed. Within three minutes we had placed a second hive body above the first and the bees began to go in again. Once more we were covered with crawling bees for neither of us had taken time to put on veils or gloves. In ten more minutes the remaining thousands had all gone in too. The field bees began to come and go. The task of making the hive a home had begun.

Neither of us was stung nor had we heard any sound of the high-pitched whine of wings that denotes an angry bee. But it is a truly spooky feeling to be covered with bees, some crawling around on one's forehead, eyebrows, lips and chin, with others investigating one's ears and neck.

There is one difficulty in acquiring bees in this way. They may or may not build straight combs from the strip starter. Sometimes bees build kitty-corner, crosswise, half-circle—or even straight combs as we desire them to do. Therefore in six days we must look in to see what they are doing. If we open the hive to look in sooner, the bees may become offended and leave. If they are building straight combs we rejoice. But if they are building any other kind, we take out and replace all the frames they have not yet started to draw with frames supplied with full sheets of wired foundation. When these latter foundation frames are well drawn and filled in three weeks to a month, we try to replace the bees' crooked starter combs with wired foundation frames also. We cannot always do this immediately because they may have eggs or young brood in these misshapen combs. Then we have to wait until the fall countdown and diminishing of bees, or until there is no brood in the crooked combs. We must remove all the oddshaped combs because the bee inspector needs to remove and examine all of the frames with relative ease. Removing the crooked combs may be a simple matter or at other times a really knotty problem. Nonetheless, the aforementioned method is sometimes one way to get bees if all others fail. Such an adventure is quite an education in beekeeping whether successful or not. May you all have as much enjoyment and good fortune as we.

Bees do have stingers and occasionally we do get stung, the fault usually being ours more than that of the bees. In the event of a sting the detached stinger should be quickly scraped off with a fingernail or knife blade to minimize the amount of poison inserted into the wound. At the present time I pay little attention to a bee sting as I am almost immune to the poison. I rarely get stung except when I accidentally squeeze a bee between my fingers while working rapidly to remove the frames of honey from a super. Old-time remedies such as mud packs, wet clay or soda packs applied to the sting area are useless in my experience. However in this era of advanced technology there is no reason for anyone to suffer the agony of a sting from a bee, yellow jacket, or mosquito. Diamond International's "Beekeeper's Supplies" catalogue lists "Sting Kill," relief from bee stings, mosquito bites, within one minute (the company does not guarantee results): Box of ten swabs, Cat. No. A-407, shipping weight eight ounces, price $1.50 plus tax. For years we have rubbed on Mrs. Stewart's Liquid Bluing available from any grocery store as this helps greatly to relieve the pain for most people. If a period of time has elapsed since the sting, rub on several liberal applications of bluing to the whole swollen area.

One day a young beekeeper stopped by to see me and approaching me like an angry bee asked, "Does a bee sting hurt you as much as it does me? Tell me the truth now for I want to know!" As he spoke he showed me the inner side of his left forearm already badly swollen and hot with fever from two recent stings. A glance at his face showed that he was almost ready to give up beekeeping.

"Oh, no." I told him. "Stings don't hurt me like that! They did fifty years ago but not any more, and in a few years they won't hurt you so much either. Let me apply some bee sting remedy." And I swabbed his arm with bluing, giving him almost immediate relief. He kept his bees.

Some years ago a baldheaded man came to visit me. It was at the height of the honey flow and hundreds of bees were coming and going from a large hive we were observing. In spite of my repeated warnings to keep back from the entrance, my visitor moved up closer and closer, then stooped down and breathed right into the entrance.

"Get back!" I warned.

"But I just want to see what they are doing," he replied.

"You're going to get stung," I told him.

"But I'm not through looking yet," he stated.

I was really surprised how patient the bees were with him—but there was a limit. Suddenly one of them stung him right on the top of his shiny bald head.

"Let me rub some bluing on the spot," I urged.

"No, no," he said, "I don't want to be spotted up."

We talked for a few minutes longer. Then he said that his head really hurt.

"Will you let me get the bluing?" I asked.

"Yes, by all means!" he exclaimed.

I did, and believe me I put on plenty. I had to because he had waited so long that his head had begun to swell. In a few minutes he said the pain had almost ceased, but he was a marked man with the top of his head painted blue. Mrs. Stewart had really put her brand on him!

In addition to liquid bluing for bee stings the following have been recommended to me: rubbing common table salt into the wound helps some people, swabbing the spot with Clorox helps others, some few are helped by Campho-Phenique. A friend recently gave me this remedy: Mix a generous dash of ordinary meat tenderizer with a tablespoonful of water and rub it on mosquito bites and bee stings as the tenderizer contains an enzyme that promptly neutralizes such irritants.

6

Beautiful Bees Waiting To Be Hived

We have let it be known that we want swarms of bees. Someone phones that a swarm is located at a certain address. If possible, we get the phone number and talk to the people living there to find out where the swarm is located—on a bush, tree, post, or wherever, and how high it is from the ground. With this information in mind we quickly load our car or station wagon with the necessary gear, usually a wooden apple box, stepladder, a two foot square of cardboard or plywood, a piece of wide board, a few short narrow lengths of board, our smoker, veil and gloves, and a properly prepared beehive. Speed on our part at this point is essential, as a swarm of bees may remain clustered on a bush or other object for as short as twenty minutes or as long as several weeks. Time is of the essence. As a rule bees do not remain clustered in any one location too long so we make a diligent effort to reach them as soon as possible. The length of their stay depends on how soon the scout bees return with information regarding the finding of a new homesite. We want to hive the swarm before these scouts return. We offer the queen a mansion—our beehive.

Most swarms are in size and shape much like a football and are usually not more than six feet above the ground. However, at whatever height they are, we build a platform under them just high enough so that when the hive is placed on the platform with the hive cover off, the hive will be just below the cluster of bees. We then give the branch or branches on which the bees are clustered a good hard sudden shake. Masses of bees will fall directly onto the frames of the hive and begin to sink down between the frames. Actually they are crawling down and are delighted to do so for they have found a new home.

Sometimes thousands of bees take wing and fly all around us after we have given the big shake. This presents no problem. We pay no attention to them as yet but continue to watch the bees sink between the frames. As soon as the hive is somewhat cleared of bees on one side, we take the bee brush and sweep the balance gently away from that side and out over the frames. Then we begin to replace the cover holding it on a level with the hive and gentle nudge it across the top until it is completely in place again. This will take about ten minutes from the time we shook the bees. Considerable care must be used to avoid crushing some bees while replacing the cover.

Covers are made in different ways. If the hive purchased has a cover that cannot be slipped on in this way, we cut a piece of quarter-inch plywood the size of the top of the hive and use it as a temporary cover.

Now as we look at the hive entrance we see many bees fanning with their wings. This is good. They are giving the "We've found a home!" call with their wings to attract the attention of all the bees flying around, and in fifteen or twenty minutes more they will be almost all in the hive. At times a mass of bees may re-cluster on the branch from which they were shaken. If so, we shake them off again. If we can get permission to do so, we cut off the limb where the bees first clustered and carry it away a short distance. Then the bees will hasten to go into the hive. We always hive a swarm of bees with the full width of the entrance open so that they may have easy access to the interior of the hive. But as soon as we have slipped the screen closure into the entrance and taken the bees home, we remove the screen and reduce the width of the entrance with short blocks of wood at each end of the opening or with a previously prepared entrance cleat. At this time a doorway for the bees only one-third the size of the entrance is sufficient, for a new swarm needs to conserve heat so that the bees can draw out honeycomb more rapidly. If the weather is very warm or the swarm exceptionally large we do not need to restrict the entrance until evening or until the weather cools.

Entrance Cleats

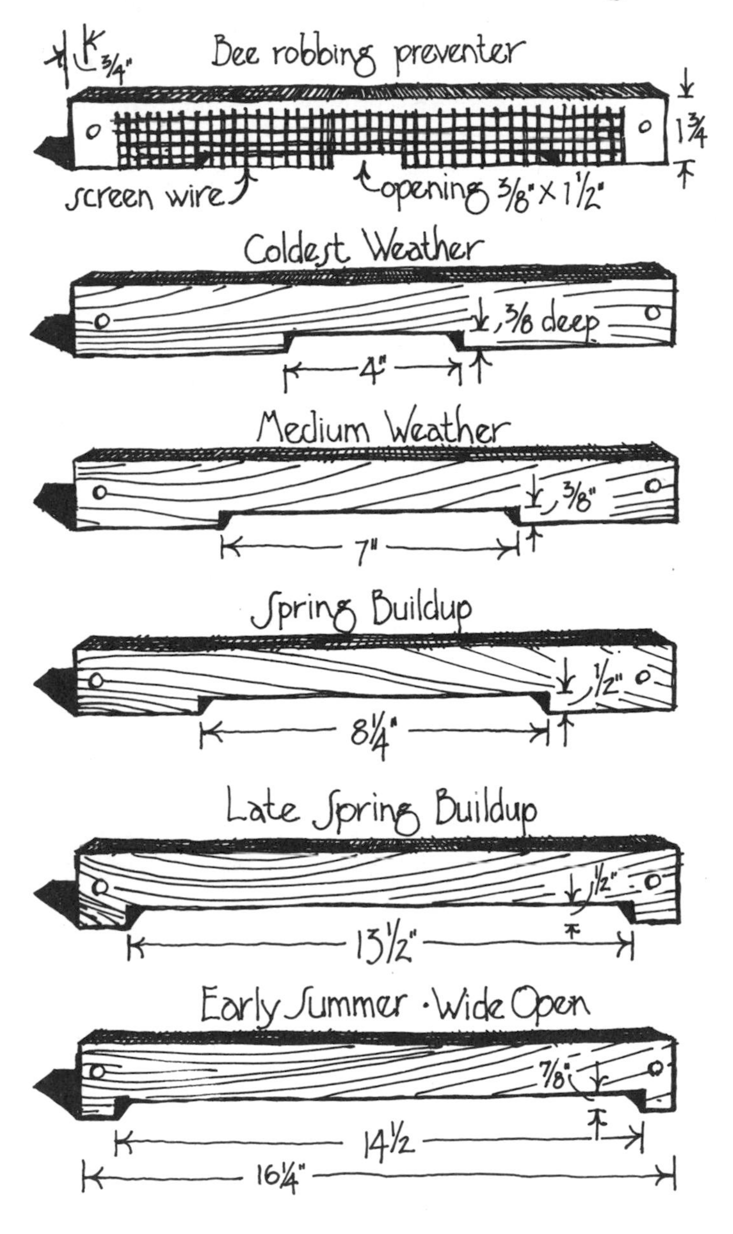

We may or may not see the queen bee go into the hive. She is an unusually long and graceful bee. If we see her enter the hive, fine; if not, no matter, for if all the bees go in, she is in the hive also because the bees will not go in and stay in without her. In a few minutes they will come out again if she is not present. We must always remember to move the hive a few feet after hiving a swarm if it is to remain until evening. This is because the queen has sent out scouts to look for a new home. Sometimes these scouts return and lure the queen and all the bees out of their nice new home and when we go for them in the evening, our hive is empty.

When almost all the bees are inside, it is time to move the hive about six feet. If the hive is on a box, saw horses or a stepladder we take it down and set it on the ground and a little to one side of where it was previously. If the hive is to be taken right home in daylight, we slip the wire screen closure in the entrance when we see as many bees leaving the hive as are coming in. We make this wire screen closure from a piece of window screen cut six inches wide by fourteen and five-eighths inches long bent its full length into the shape of a "V" so that when placed into the hive entrance its own spring tension holds it securely in place. On occasion, we use another type of closure, a full-width entrance cleat with wire screen securely tacked over the cleat's opening to permit the free circulation of air during the time the bees are penned in the hive. With the bees secured, we load the hive into our car and go home. If a few stray bees enter the car as we prepare to leave, we pay no attention to them for they will not sting but will go to one of the windows and try to escape and will continue doing so until we reach home and help them out. This is easily done by placing the open end of a water glass over them and slipping a piece of stiff paper between the window and the glass. When the bees are caught we can look at them if we wish or let them go free at once.

If a hive is taken home in daylight some dozens of bees will be left behind and this cannot be avoided. They will usually return in a day or two from where they came. If all the bees are wanted, or it is required by the landowner that this be done, then the hive must remain in its new location until just after dark at which time the wire screen may be placed in the entrance and the hive loaded and taken home. Always remember to remove the screen closure upon arrival at home.

Caution: Before moving a hive with bees in it, always tie a rope or wire tightly *around* the hive, over and under it, so that the top and bottom are securely held in place. If either the top or bottom becomes displaced while the hive is being carried, one will surely be stung. Old hives sometimes have holes other than the entrance. These should be plugged with soft clay or putty.

Bees rarely sting while being hived and there is little need to be afraid. However, it is well for most people to wear a screen veil and gloves just to be on the safe side. We pay no attention to an old beekeeper who may be present and give us the horse laugh for wearing a veil and gloves. After all, if trouble should develop he can run away, but the beekeeper must stay and finish his hiving. The rule is that bees do not sting during swarming and may be hived easily. There are exceptions to this rule. If a swarm has been unable to find a new home, it may remain clustered on a post or among vines for as long as two weeks. Cold or rainy weather also causes bees to remain clustered until they become hungry; they are cross when disturbed, so play it safe. As time goes by, many folk will learn to hive bees without veil or gloves as some of us do, but always have the veil and gloves handy.

Now and then bees cluster in a location where it is difficult to build a platform under them. In such situations we find it easier to hive the bees with the hive placed on a piece of cardboard on the ground. The cardboard is used to keep the bees from getting lost in the grass or leaves and also to keep them from going *under* the hive instead of *into* it. The side of a refrigerator shipping carton makes a good base for this purpose. We use the little boards we have brought along to close any cracks remaining between the cardboard and the hive. Next, with the owner's consent, we carefully saw off the branch from which the bees hang suspended and carry them over the hive and shake them onto it as already described.

It is difficult to hive a swarm that is clustered more than eight feet above the ground so a beginning beekeeper should hive some lower swarms first. If one is notified of a high swarm he does not feel able to handle, he should tell the people who notified him of this fact so that they can call a more experienced beekeeper. This always makes for good public relations and that is most important to us as beekeepers because when one has bees he soon has honey to sell, and we all need buyers for our product.

Hiving a swarm of bees from a fence post is quite difficult, yet we often meet with this situation. We have tried several methods and the following has worked out best. For example, let us say that a swarm of bees is found clustered on a five foot high fence post partially covered with berry vines. We may not cut any of the vines or remove the post. We take a pair of saw horses, carefully part the berry vines, and place one saw horse with the middle of its back within six inches of the post. We place the second saw horse eighteen inches farther away in the same position as the first. This makes the platform. We place the hive on the saw horses with the entrance end of the bottom board *touching* the post on as wide a surface as possible. The bottom board thus forms a bridge between the post and the hive for the bees to cross as they enter the hive. We always take along a wide board to use in case there are so many vines that we cannot push the hive entrance up against the post. The wide board serves as an extension to the bottom board. In the above example bees always crawl up or down the post to the entrance board and on across it into the hive. They will refuse to fly across what we may think is an insignificant space between the post and landing board. For this reason the hive bottom or extension board must touch the post to encourage the bees to enter the hive.

Often the bees completely cover the post and vines from ground level to the very top. The post cannot be shaken so in lieu of that we take a colander or dipper and dip up a quart or so of bees from wherever they are the thickest and dump them on the bottom board in front of the entrance. They will immediately set up their fanning call and attract many more bees to come to the hive. If possible, we scoop up several more dippers full and empty them in front also. It is impossible to tell where the queen is in this situation. If she is near the entrance she will go right in and soon all the other bees will follow her. However, if she is near the ground or on top of the post among the vines she may not come up or go down. Then the bees we have emptied in front of the entrance may begin to leave. If this happens we try to empty some more bees in front of the entrance. We start up our smoker and give the bees on the ground six or eight good puffs to start them climbing up. We do the same to those on top of the post to start them crawling down. As the smoke clears we repeat the process as soon as the bees stop moving. Patience and perseverance usually pay off and in an hour's time we have our bees.

Our most important tool when working with bees is a good smoker, a mechanical device with a hand-operated bellows attached to a four inch diameter, ten inch long can-like compartment with a metal grate at the bottom and a removable hinged spout at the top. We cut up old burlap sacks into eight inch square pieces and use them for fuel in our smoker. We open the smoker, hold a piece of burlap over the open end and set fire to it with a wooden kitchen match. As the burlap takes fire we press it down toward the bottom of the smoker until the burning material is just above the grate, at the same time giving the flame a few light puffs of air from the bellows. As soon as the first piece of burlap is burning well we add more pieces until the smoker barrel is filled. At the same time we continue to work the bellows. Then we close the lid so that most of the smoke generated will come out of the spout so that we can direct it where needed. We always use burlap as fuel as it is readily available from feed stores and the smoke is not harmful to our bees or to us. Some materials, such as plastic cloth or drape fabrics when used as smoker fuel, emit noxious fumes with the smoke which if inhaled is sickening if not truly deadly to both bees and beekeeper. Burlap smoke when applied to our own hands, feet, and face helps to eliminate our body scent causing the bees to be more docile as we work with them. When applied to the entrance of the hive it makes the bees think their home may be on fire and the guards and all nearby bees attempt to solve that problem and pay less attention to what we are doing to their hive. I find it a good rule of thumb to use the smoker more to smoke me and my partner than to smoke our bees, though at times both are necessary.

"Look before you leap" is an old cliché but never more true than in hiving swarms of bees. It always pays to stand to one side of the swarm for a few minutes and observe carefully the bees to be hived, their position on the bush or branch, and what the immediate result will be of taking the steps necessary in hiving them.

Two friends of mine, Ronald Lang and Kenneth Gray, recently reminded me of this fact. They suggested that I relate their recent experience as it could be of help to others. They found a large swarm hanging from a higher limb of a big willow tree. Below the swarm was another long graceful limb which was in the way. Standing on the ground directly beneath the swarm without his veil on, Ronald pulled down the supple obstructing limb and cut it off with a hand-

Smoker

saw. Swoosh. Up flew the remaining portion of the severed limb, now released from the weight of its heavy lower growth, and struck the swarm of bees a resounding blow squarely in the middle. Down came the bees in a tremendous shower—all over Ronald—from head to foot. Being frightened at such rough handling the bees stung and stung—forty or fifty times in a matter of seconds. They were in his hair and down his neck. He ran for home with Kenneth running alongside brushing off as many bees as he could and smashing others. Ronald began to have a serious and then critical reaction to so many stings. They loaded him into a car and Kenneth burned rubber for the nearest hospital. Before he arrived there Ronald had retched and lost consciousness. But they got to the hospital in the nick of time. The doctors administered oxygen, a heart stimulant, and whatever else they do in such cases, and Ronald recovered.

After this near fatal experience, Ronald thought he would get rid of the two hives he already owned and quit for good. Then he thought it over, realized it was his own fault that he had been stung so badly, and decided to keep his bees. When he returned home from the hospital he and his friend Kenneth went to see what had become of the swarm. They found the bees in a heap on the ground where Ronald had been standing. So Ronald urged his friend to come to see me, buy a hive, and hive those poor bees. This he did—and that got Kenneth started with bees—a fine hive as I observed a few days ago. Mrs. Lang said she was pleased her husband had decided to continue beekeeping in spite of his ordeal because ever since they had owned their own bees and had been eating their own honey she had not had any attacks of hay fever such as had caused her so much misery every spring in prior years.

During my lifetime I have noticed that some of the largest and finest swarms cluster on a branch eighteen or twenty feet above the ground, so far out near the tip of a long limb that one cannot build a platform in the tree upon which to set a hive. The following instructions are tricky but I have used them many times. I get an extension ladder and raise it and place it so that it is just a little toward the tree side of the swarm hanging from the branch. The ladder is placed almost straight up because the branch of the tree up there may be only about one inch in diameter and shaky. If I have a bee partner with a really good set of nerves, I have him put on his veil and steady the ladder. Then with my own veil on I climb the ladder and tie it to the limb with a short rope as a safety measure. I lean toward the swarm of bees and blow my breath over them through the veil and talk to them. If they pay no attention, they are gentle and I may go ahead and have no fear.

Back on the ground again, I take off my gloves and roll my sleeves up to the shoulders. Then with a sharp saw in one hand I climb the ladder and gently saw off the limb, holding it firmly with the right hand, sawing with the left. In spite of all I can do the weight of bees will force the branch from off an even keel as I continue sawing. This will cause a mass of bees to be dislodged from the lower end of the swarm and they will fall almost to the ground before taking wing. Up they will come with a tremendous buzzing—but they mean no harm. They merely want to re-cluster so I give them a couple of minutes to do so. This gets to be hard work, for one is standing with one foot on a ladder rung and the other leg hooked over the next higher rung to keep in balance while sawing. I lay aside (sometimes have to drop) the saw as soon as possible and grasp the sawed off branch with both hands for a little while to rest my right hand as the bees re-cluster. The end of the limb with the bees is now hanging lower than my hands. Bees always want to climb upwards so in a few minutes they start to cross the few inches of bark between my hands and the swarm. Moments later they begin to cross my bare fingers and climb my bare arms. This is a bit scary but there is no danger because with sleeves rolled up the bees cannot be squeezed between arm and cuff and therefore have no reason to sting.

When most of the bees have re-clustered, I hold them again in my right hand and begin to descend the ladder. It takes careful handling not to shake the swarm unduly, for if I do, some of the bees may again become dislodged from the cluster and fall upon my partner below before they can take wing. I take special care not to place my left hand on a bee resting on a rung of the ladder. I always descend slowly and when safely down carry the swarm to wherever the hive is located even if it is as much as 200 feet away.

People who have seen me bring down a swarm of bees in this way for the first time have thought that I had lost my mind for sure! However, after seeing it done a few times some have considered doing it themselves. I have never been stung even once bringing down a swarm in this way. I am pleased to say that others have been as successful as I in bringing down such swarms.

It is my partner on the ground holding the ladder who deserves much of the credit for my success. A good bee partner is one of a beekeeper's greatest assets. In every neighborhood there is a boy or girl who makes an excellent bee partner. Gary Blankenbiller was sometimes my partner in difficult situations. He has real nerve on the ground. Also he showed outstanding courage one day while we were hiving a swarm of bees high up in an oak tree.

Late on a warm afternoon in May, Gary and I went down to the oak woods at the lower end of our property to see if we could find any new swarms of bees that might have clustered there during the earlier hours of the day. Sure enough, there we found a large swarm away up in the very tip-top of one of the taller oak trees that fringed the woodland.

"Are you going to hive it?" Gary asked excitedly.

I took a good look at the way in which the swarm was situated in the tree. Most woodland border trees have a nice leafy green rounded aspect, as did this tree the exception being that out of the very center of the crown one branch had grown straight up for about three feet. The bees were clustered on this vertical branch. There was no way we could cut off the branch and carry the bees down still clustered to it. If we were to hive this swarm we would have to climb the tree, build a plank platform under the bees, place a full size hive body on the plank, and then brush and coax the bees into the hive.

It was a beautiful swarm of leather colored *Italians*. Common sense ruled out any attempt to hive such a high, strangely situated swarm. As I continued to gaze up at the bees a dread, almost akin to fear, swept over me.

"Are you going to hive those bees?" Gary again asked eagerly.

"No, I'm not." I answered. "It's entirely too risky. My father and I don't need those bees."

Then to my surprise Gary urged quickly, "Hive them for me! I have an empty hive all ready and waiting. They're such beautiful bees. We really could hive them, couldn't we? You do know how? Let's get them for me."

Such abounding faith, hope, trust, and eager expectancy young manhood exhibits! I took another searching look at our high swarm and yielded a trifle.

"It will be not only a tricky maneuver but also positively dangerous," I warned Gary.

"But you admit that we *can* do it?"

"Yes."

"Then I'll go get my beehive," and away he went.

"Bring the longest ladder you can find, too." I called after him.

During the following ten minutes I had time to think, as I gathered together the hiving equipment we would need. The operation would be dangerous. I pondered what I could do to minimize the danger. The answer was to take along a stout rope and also get a positive promise from my partner that he would climb the tree to work with me and stay beside me in any emergency no matter how badly we might be stung or whatever else might happen.

When Gary returned with his hive and a ladder I asked him if he was ready and willing to climb up to the swarm and stay with me until we were both safely down again. As he looked up at the bees he considered the matter carefully and then agreed.

In a matter of minutes we had the ladder set up against the tree. It reached up to the first large branches. Gary climbed up ahead of me with a handsaw and cleared away the intervening small branches so that I could follow bringing the hive and a ten foot long new plank to serve as a platform for the hive to rest upon.

All went well. We placed the plank and hive in an advantageous position and tied everything in place with the rope—that is, everything except me. There was such a handy stout limb for me to stand on that I saw no reason to tie myself to the tree as we sometimes did. We hived the swarm in twenty minutes as the bees were most cooperative. The entrance to the hive was out of my reach so I decided to take the hive down without placing a screen in the entrance to confine the bees.

I told Gary to start climbing down as I began to pull the hive toward me along the smooth plank. It would not slide. I pulled harder. Still it would not budge. I looked carefully along each side but could see no reason why the hive would not slide along the plank toward me. I gave it a hard jerk. Instantly it slid toward me a full two feet. It was still resting on the plank but it had slipped so far that I was off balance. I made a desperate grab for the nearest limb. My fingers pressed the bark on top of the branch for a few seconds and then slipped off. There was no other branch within reach. Instantly I grasped the beehive again which was now resting with its back end against my chest.

"Gary," I cried out in a low voice, "I'm overbalanced backward and falling."

"I'm far out of reach but coming!" he shot back.

As though by a miracle the beehive caught on the plank and my fall was stayed. But a few seconds later it slipped free again and slid toward me another inch. I held on desperately, pressing down on the hive as hard as I could to give it more friction on the plank. With increasing unpredictability the hive would stop, then start again to slip toward me and my sense of falling became more acute. The vibration annoyed the bees. They began leaving the hive to circle around in ever increasing numbers.

Thoughts flew through my mind. I had disobeyed both safety rules. I had not tied myself to the tree nor had I kept my partner by me. I glanced down. There was no hope of survival if I fell from that height, and in addition, I clutched a heavy beehive filled with escaping bees tightly to my chest.

I could hear Gary coming.

A sudden shift of the beehive ended all thinking. I had begun the final plunge. In that split second Gary's strong arm threw a circle of salvation around my shoulders.

"Got you!" he cried exultantly into my ear. No man was ever more thankful to be "got" than I was at that moment. Slowly he forced me back into an upright position onto the limb on which I had been standing. The beehive was still in my arms. After we had rested a little I lowered it down to the ground with the rope. Gary had his coveted bees.

Next day he came over, put his arm around me and said, "Ormond, no matter what I say in the future, don't ever again take such a chance for me."

I nodded in agreement. The adventure had been worth the risk. We had both learned a good lesson in beekeeping.

Through the years the hiving of new swarms has brought me into many other interesting situations. Until our hives are all full, my father and I are almost always ready to go hive a swarm of bees at a moment's notice. There is a challenge in the operation that a pair of old beekeepers with a true case of bee fever find irresistible! Some years ago I received an urgent telephone call from the owner of a restaurant in downtown Santa Cruz. He said that a large swarm of bees had just clustered on his big signboard high up over the door of his establishment, and that they were stinging his customers. He asked if we would please hurry down and *do something*!

In a few minutes we had loaded our ladders and other gear and were on our way. On arrival we placed a ladder and I climbed up and checked out the bees by blowing my breath over them. Not even one bee gave out with the angry call. I went down and asked the owner how many people had been stung and he admitted that as a matter of fact no one had been stung *yet* but that he was afraid that any minute someone *might* be. I nodded, for that is the usual answer, but I did agree with him that it could happen since there were indeed many people passing by under his sign and there were still several hundred bees buzzing above his doorway.

A second glance upward showed us that it would take a bit of clever maneuvering on our part or someone might in truth be stung for the bees were clustered on the face of the sign all around and behind the neon light tubes. We quickly raised another ladder, improvised a platform on top, carried up an empty hive, and began brushing bees in toward the entrance. Fortunately for us, these bees were thankful to find a new home and in fifteen minutes the greater part of them were safely inside the hive. I placed the screen closure in the entrance with a feeling of genuine relief. A few stray bees flying around alighted upon the extended bottom board as we prepared to lower the hive to the sidewalk, but such bees never sting. I descended the ladder and waited on the sidewalk as my father lowered the hive with a rope. As I was untying it a heavyset young man stepped close to me, apparently unafraid, to see what I was doing. The hive was awkward to load into our station wagon alone so I looked up and said to him, "Say, how about giving me a hand getting these bees loaded?"

A strange expression spread over his face and he almost shouted, "Not me, I'm not brave!"

Then he jumped backwards—a tremendous leap—straight back! Fortunately everyone else was watching from a distance or someone would surely have been hurt, not by the bees, but from being knocked down by our absentminded, overly curious spectator. He created a new Olympic event and set the first world's record for the reverse broad jump.

7

New Swarms, Combining Swarms

NEW SWARMS that we put into new hives and expect to keep for ourselves or sell to others seldom make us or the new beekeeper any honey the first season. This is especially true in those parts of the world such as ours where the bees are solely dependent upon wild flower bloom for all their pollen and nectar gathering. In cultivated areas where pollen and nectar-bearing crops are grown, bees may produce excess honey because the honey flow covers a more extended period of time. Our new hives seldom make excess honey the first year because our weather is cool and it takes more time for the bees to make new wax, draw out the foundation combs, gather nectar and pollen, and start broodrearing. Furthermore, the honey flow in many wild flower areas has practically ceased by the time a new hive can build up enough in strength to produce a surplus.

There is also a great variation in the size of swarms that come from different hives. Some hives that have wintered exceptionally well build up rapidly and send forth a large swarm of ten pounds weight (about 40,000 bees) or even more. Some of our largest swarms do not come from beehives at all but from unused water tanks, huge hollow trees, or the walls and attics of homes. These swarms are the ones we rejoice to catch as they often *do* make surplus honey the first season. These very large swarms begin drawing out the starter comb the same day they are hived and make rapid progress. It is a real joy to watch a large swarm take up housekeeping. Nevertheless, until all of our hives and equipment are full and in use, we are always glad to hive whatever size swarm we can get especially if it is near the beginning of the bee season.

Within a short time after hiving, even a modest-sized swarm will begin to draw some comb and the queen begins to lay worker eggs in the partially drawn cells. But even so, it will take twenty-one days before these eggs become young worker bees. These young bees usually spend their first five days or more as nurse bees in the hive, so that it will be about one month before young bees are added to the field force. The queen must have a large field force in order to gather enough stores to supply the hive for the winter.

Here is a comparative table of the normal duration of the bees' transformation from eggs to winged insects.

	Days		
	Queen	Worker	Drone
Eggs	3	3	3
Growth of larva	5½	6	6½
Spinning of cocoon	1	2	1½
Period of rest	2	2	3
Metamorphosis into pupa	1	1	1
Duration of this stage	3½	7	9
Average time from egg to winged insect	16	21	24

Combining Swarms

Quite often we are called to hive a swarm that is really too small to gather enough stores to survive the next winter by itself. Of course, we can feed them beginning about September first and thus carry them over, particularly if we desire to increase our number of hives. But let us say that we already have bees in all of our hives, although some of them are weak. Weak means short of bees per hive. Someone phones us to come get another swarm. We would like to if we had another empty hive, but one is not available. We proceed as follows. We take an empty super and place ten frames in it. These frames have wax starter sheets only. We give the super a hive body bottom with entrance opening and a hive body top. The result is a small beehive. Now we hive the swarm into this super in the usual way and take it home.

Immediately upon arrival at home we must begin combining these bees with a hive we already own. This may be accomplished in various ways. We pick a weak hive and, after giving it a few puffs of smoke at the entrance from our smoker, remove the top, then lightly smoke and brush the bees clinging to the top cover back onto the frames. Over the top of the open hive we place a piece of window screen tacked to a wooden frame built the same size as the top of the hive so that no bees from below can mingle with those above. We then remove the *bottom* from the newly acquired bees and place them gently on top of the screened established hive. Bees will fight and kill each other if placed directly together with no screen between. We leave the screen in for twenty-four hours, removing it next day using smoke and gentleness. By that time the new bees will have received the hive scent from the established hive and the bees will not fight but will work together in harmony.

Ordinarily there would not yet be a queen excluder on such a weak hive but if there is one on, we must be sure to remove it. We replace the top super with the new bees on top of the established hive and remove the cover, then gently smoke and brush the bees from each frame down onto the hive body frames below. We remove the now empty super from the hive and replace the cover. The hive is again as it was before combining but with a greatly increased work force.

When beginning the final combining operation as noted above after the twenty-four hour screen separation waiting period has been completed, we pry the hives apart with a hive tool *below* the window screen frame. We raise the upper part but not enough to let any bees escape and then smoke the bees quite thoroughly on all sides through this small crack below the screen frame. This must be done to drive the established lower hive bees down and at the same time start the new swarm bees climbing up to the area under the cover. Sometimes the two queens are glaring at each other eye to eye through the screen until the smoke drives them apart. Now we have our partner hold the upper part of the hive (or I set it to one side leaning it against an adjoining hive if working alone) and quickly remove the screen frame, brushing any adhering bees rapidly but gently onto the frames below. We replace the upper part of the hive on the established hive and proceed as already described.

Caution: Never try to combine using a screen without a frame, thinking to have your partner lift the upper portion a trifle while you pull out the screen. Chances are you will roll and crush or injure *both* queens and lose the hive altogether.

If both queens are killed outright, the established hive bees will immediately realize that they are queenless and will begin to rear a new queen from a worker egg or day old larva. This causes a serious delay for the bees and beekeeper but is not fatal to the hive as a whole. However, if both queens are only injured they will stop laying eggs and it may be some days before they finally succumb. In this sad situation it is my belief that the established hive bees let the eggs all hatch and the larvae become too old for queen rearing purposes before they realize they are going to be queenless. Then the bees are in a desperate and hopeless predicament. So is the beekeeper if he has no other queen bee available. It is distressing to see our hive of bees slowly diminish in numbers, and after two months or so die out completely. Actually, we dare not wait until the last bees die off but must open the hive and drive them out because if we delay too long the wax moths will take possession of the hive and destroy all the combs previously drawn and filled with honey and pollen.

Caution: Always remember to remove the screen the next day or within forty-eight hours at the latest. Never let the new bees remain penned up to starve!

As a new beekeeper gains experience, he may want to use the following method but he should have a helper to assist him. We bring the new bees home and place them just as they are on top of the hive we want to combine, remove the screen closure to let them fly freely for the rest of that day, and wait until *after dark*. Then without any light we very gently set the top hive to one side, or better yet, have our partner hold it. Very gently and without any smoke or other disturbance we remove the top cover from the established hive and set it to one side. We then remove the bottom board from the new hive and set it also aside. We place an empty "in between" super on the established hive, then carefully lower the new bottomless hive with its bees down onto the "in between" super with *no screen* separating the two swarms. In place of the screen to keep the bees from fighting and killing each other, we have provided a relatively large open air space in the hive to separate the bees until the new swarm can acquire the hive scent.

Caution: Always stay close by and watch for an hour with the smoker going. If the bees begin to fight at the entrance (always have a flashlight handy) smoke them thoroughly at the entrance, raise the cover slightly and smoke them heavily there too. Smoke them thoroughly several times and they will stop fighting. It is truly disheartening to find thousands of dead bees clogging the entrance next morning due to the fact that we did not keep watch long enough the evening before.

The advantage of this method, tricky as it is, is that we do not have to open the hive and remove the screen the next day to release the bees. It is particularly useful if it appears the weather is going to be cold or rainy next day.

There is one other problem when using this method but I think we have found a good solution. If the swarm is hived early in the day it should, as soon as is practicable, be placed on top of the hive to be combined. By day's end many of the new bees serving as fielders will have located their hive entrance as being one foot higher than it will be after combining has been completed the next day, causing many of them to act lost for hours. To remedy this situation, we place a two-foot square piece of cardboard just above the entrance to the upper hive and tie it in place with a stout string or wire. All that first day the bees locate themselves by the cardboard as well as by other features of the landscape. When we combine hives the next day and lower the cardboard to just above the entrance to the lower hive, the bees follow the cardboard down. They find the entrance far more easily and with less loss of time when we use a cardboard roadsign.

There are people who stoutly affirm that bees have a remarkable homing instinct. This may, in part, be true but experiments and observations my father and I have made would indicate honeybees primarily make more intelligent and even greater use of the five senses than do human beings, when seeking sources of nectar and pollen out in the fields or returning loaded to their hives. They may also surpass us in the use of Alpha, Beta, Theta and Delta waves.

In our experience, after the hives have been combined and the queens fight it out as they always do, the *newly added* queen will be the victor in the battle of queens, and thus be the surviving queen of the hive. The rule is that there are never two queens in one hive. However, in our many years of beekeeping we have had one exception. It happened in this way. All our hives were full and strong. Then a new swarm came from somewhere other than our own apiary and settled in an oak tree nearby. We decided to catch and combine this swarm with our No. 6 hive according to my first method given. We had hived No. 6 five weeks earlier and they were ready for a queen excluder and their first super. Now we did not do that. We hived the big new swarm in a full depth super (standard hive body) and combined it with No. 6, a hive of black bees, Caucasians they are called. The new swarm consisted of light three-band Italians. Several months went by. We put on a queen excluder and supers as required. They made us three fine supers of excess honey.

Then the bee inspector came around. "What is this?" he inquired, "Caucasians and Italians in the same hive in late summer working together?" They surely were working together and had been all summer long. Apparently, in combining, each queen had stayed in her part of the combined hive and kept house as a good queen should. I do not know how much longer the queens kept house together for I sold that hive. But they worked and lived in harmony for one full season anyway.

8

Enemies Of Bees

ONE OF OUR GREATEST PROBLEMS is how to control the tiny, spindly, brown ants that we find in countless numbers in this part of California. During the summer and fall their nests may be almost anywhere—under a big stone, piece of board or flat metal, near a tree, grape vine, or fence post. They try to come into our houses to eat and drink with us as well as infest our beehives. How to live in peace and harmony with them is a continuing problem.

Poison is not the best answer. Most people use poison for lack of a better way to remove ants from their premises. As beekeepers in a wild flower area we use no poison dusts or sprays of any kind lest we kill our bees. During most of the year the ants are our friends. They do a tremendous job of helping keep our apiary and yard areas clean. During the height of the honey flow our field bees work so hard that they live for only six to eight weeks. If they die in a hive, the hive bees carry them out and often fly away a short distance before letting them drop to the ground, or before alighting on the ground with the dead bee. Sometimes the live bee has quite a tussle disengaging its hooks from the dead bee and can free itself of its lifeless burden only when the dead bee becomes wedged in the grass so that the live bee can finally pull itself free. In any event as soon as the carcass is abandoned, the ants strip it of all edible flesh.

Many years ago I began asking people how to rid a beehive of ants without using poison. I received numerous seemingly quite logical answers. One senior citizen told me to cut off all four legs of my hive-stand at ground level, set my stand with its bees aside for the moment, and then nail a pie pan to the top of each sawed off post. Then I should replace the stand with its legs centered in the pie pans and fill the pans with oil. I did as directed. It worked fine. The ants could not cross the oil so the bees lived in peace. Then came a heavy rainstorm. The pans filled with water and the oil floated off. I was not too badly dismayed because the ants could not cross the water barrier. I went on my way rejoicing.

A few days later I looked at my hives again and found them overrun with ants. You know what had happened. After the rain the sun and wind had evaporated the water from the pans and the ants had crossed like the Children of Israel over the Red Sea on their way to the Promised Land. Well, the ants had also regained their Promised Land—my beehives. I refilled the pans with oil. Next day we had a strong north wind. Leaves, bits of bark and trash of all sorts slithered along with the wind. The oil pans made perfect catchers' mitts for everything that passed by. Again the ants passed over on dozens of natural bridges. I cleaned out the pans, put in fresh motor oil, and tried again.

For some weeks all went well. Then one day we had a strong wind from the east. Toward evening I went out to look at my hives. With consternation I saw that during the day many of the heavily laden homecoming bees had been blown off course right at the hive entrances and had fallen down into the oil pans until the pans were all full of dead and dying, oil-soaked bees. A close inspection revealed ants beginning to cross over the oil on the bodies of the dead. Then I saw something that *really* shook me. Among the many dead bees was the body of a beautiful fertile young queen, a leather-colored three-band Italian! It was the queen of the hive occupying the center position on the stand. I was shocked and dismayed. I called my father and together we got rid of those oil pans down near the ground once and for all. At that time we were not skilled in combining swarms so we lost that hive.

At first I felt a touch of bitterness toward the ants for the loss of our queen and the whole hive. On second thought I realized that our loss was due, as a matter of fact, more to my efforts as a bungling beekeeper than to the activities of the ants. The ants had not poured the deadly oil into the pans—I had. This experience pointed out an important lesson in beekeeping. We must never try to play the part of the Good Samaritan to our bees to the point where our efforts are more of a hindrance than a help.

Through the years we have continued to experiment with oil pans of various sizes, depths, and positions, under at least one of our hives. None of them has been completely successful although some have helped temporarily. This past fall my father placed a hive on four big steel nuts taken from bolts. These nuts were placed in small oil pans completely hidden under the hive body so that no bees could fall into the oil. The nuts held the bottom of the hive three-eighths inch above the rims of the pans and much higher at all other points. We watched the ants try in vain to get down from the hive. Those on the stand below tried equally unsuccessfully to get up. I thought my father had finally solved the problem this time.

A week passed. Then I noticed that once again the hive was overrun with ants. Surely a stick or tall stalk of grass must be leaning up against the hive to serve as a bridge. Not so. I got down on hands and knees and looked carefully all around under the hive. There was no sign of a bridge of any kind. Then I located the line of march of the ants and followed it to the rim of one of the oil pans situated in the darkest corner. There I saw how the intelligent little insects had solved the bridge problem. Two of the ants that had been marooned on the hive when we had placed it up on the big nuts, had reached down and grasped the "hands" of two ants reaching up from down below. Together the four ants formed a neat bridge over which one line of march ascended to our hive and another line descended. Once again we had the privilege of rethinking our former ideas.

The big nuts did not work too well at best because it was difficult to keep them centered in the little pans. This seemed to be due in part to earth tremors. Steel nuts surrounded by oil slip too readily when placed in steel pans. In our latest experiment we used four salad dressing jar lids, quart size, and four one-and-one-quarter inch long lengths of large broom handle. We centered the short lengths of broom handle in the lids and then turned them over and drove a small nail through the lids into the wood. When turned upright again, the pieces of wooden handle stayed securely in place even when surrounded by oil. This method seems to offer real hope for ant control during the fall and winter months when we have few if any supers on the hives.

I think we will have to remove the oil pans before the main honey flow begins because the broom handle supports have such a small bearing surface that they will exert an undue pressure on the hive bottoms. Large hives filled with honey are very heavy. We may try placing supporting boards from one oil pan post to the next to relieve the pressure on the hive bottoms if this experiment proves successful in keeping our hives free from ants until the honey flow begins.

Spiders are a menace to our bees particularly during the spring and summer. They like to hide under the piece of corrugated aluminum we use for a hive roof. It is surprising how many bees spiders can catch and kill if we do not frequently raise the roof covers and brush them off. Some of them in the space of a single night build huge webs of exquisite symmetry and beauty. It seems a shame to destroy these lovely creations but we are compelled to do so as they are death traps for our bees. At times I have destroyed one of these large webs every morning for seven consecutive days before the persistent old spider gave up and moved elsewhere.

California jays can put a one-hive beekeeper completely out of business. Most birds do not become bee eaters. Those few that do literally live on bees. Some years ago we hived a medium sized swarm and placed the hive in a remote corner of our property. Six feet in front of the hive was a five-foot high fence post. We watched. Soon we saw a jay alight on the post. Moments later it popped up into the air and caught a bee flying past. One big gulp and the bee was gone. Nine times in a row it caught a bee and returned to its post. The tenth time it flew away to the woods. At approximately two-hour intervals all day and every day it returned to gorge on more bees. For three weeks we watched the bird come and go. Then it failed to show up for its two o'clock feeding. Puzzled, my father and I opened the beehive. There was not a single bee left inside. That gluttonous old jay had feasted on the bees from that particular hive until there was not a bee left!

This experience proved the guilt of some jays and started our search for a way to outwit them. We found a simple and effective method. We set spring-type mouse traps on each fence post near our bees. We did not bait them. When mister jay alighted on his favorite perch he usually sprang the trap and got a resounding whack in the pants as the wire bail of the trap flipped shut. Away he would go screeching mightily. Two lessons were all most jays needed. We had a few that refused to take warning. These birds became so bold as to alight directly upon the extended landing board of a hive and at their leisure pick up and eat the hard working bees serving as fanners to ventilate the hive. For them we set rat traps baited with pieces of bread. These traps we placed on the extended landing boards of our hives. Jays love fresh bread even more than bees. I have seen a jay alight upon the business end of a rat trap and peck at the bread bait. When the trap snapped shut the powerful spring caused the wire bail to catapult the jay ten feet forward. Old mister jay usually flew away at least fifty yards before burning our ears with his scoldings. We rarely actually caught a bee-eating jay. We were pleased with this fact because we did not want to kill the birds, just persuade them to desist from eating our bees.

In a broad sense all birds, including jays, are a beekeeper's and gardener's best friends. We do everything possible to encourage birds to nest on or near our property. Sparrows search every nook and cranny and are indispensible in the control of spiders. Spiders should not be despised either as they trap in their webs multitudes of lesser insects that if left unmolested would make life a misery for us. Only those spiders that feed on honeybees are the ones we wish to control. We encourage all of the smaller birds—robins, towhees, blackbirds, mockingbirds, quail, and many others—to scratch and feed near and under our beehives. For the home gardener we have found no pest controller more effective than a couple of ducks. Snails, slugs, sow bugs, beatles, everything small and large up to and including Jerusalem crickets are king's fare for a pair of ducks.

The wood louse, called sow bugs in this country, is a pest if it decides to make a hive its home during the fall and winter. They do not do this too often, fortunately, but when they do they keep the bottom portion of the hive wet and messy with their manure and general life style. A hive of bees with a wood louse infestation does not thrive. The bees seem unable to drive them out. When I find this type of situation I separate the hive proper from its bottom and at dusk set the hive with its bees to one side. Then I scrub the bottom board thoroughly and reassemble the hive as quickly as possible. Such treatment usually ends the wood louse problem for that hive. In my latest experiment now under way, I leave the hive as it is but remove all the restricting entrance cleats so that the entrance is wide open for a few minutes as it would be if it were summertime. Then I take an ordinary handsaw and slip its blade into the hive entrance all the way back to the rear of the hive and scrape it along the back wall until I come to one side. I saw the blade back and forth to dislodge the bugs and their mess and then draw it out. Wet mess, manure and bugs all come out together on the flat saw blade. An application of this cleaning method done every other day for three days has proven effective in ridding sow bugs from several hives this winter. Of course I always do this early in the morning before any bees are up and flying. It is surely an easier way to get rid of the bugs than disassembling a hive.

Some beekeepers scatter poison bait on the ground behind the hives and cover the poison with a board to attract the sow bugs to the bait. How well this works I do not know. Personally I have not found it to be very effective.

Poisons of various kinds used as sprays or dusts around the home and garden are a menace to our bees. Some are much more toxic than others. In our experience DDT affects bees less adversely than most, possibly because bees are not attracted to it as they seem to be to some of the other poisons. In past years we made judicious use of DDT around the outside walls of our extractor room at the ground level to control ants in late summer when the ant colonies are very strong and most persistent in their efforts to overrun all our honey equipment.

Occasionally a hive of bees will slowly die out and leave the hive empty for no apparent reason other than that the queen bee may have died because of injury, old age, or poison. I am often asked if a new swarm may be hived directly into this empty hive. My answer is no. We always detach the bottom and top from such an empty hive, then remove all its frames. We build a good fire in our big oil drum incinerator and when the flames shoot over the top we scorch the bottom, hive body box, and the top, almost to the point of charring the wood.

Next we examine the frames. If we find residue of any sort in the cells of the drawn combs, we dig a hole in our garden about two feet in diameter and two feet deep. Over this we place our oil barrel incinerator. Then we take a hose and with the water at full force we blast it against the wax combs as we hold them in the barrel. The cold water strengthens the combs and at the same time washes out whatever material remains in the cells. When the combs are all cleansed we shake them dry as possible, making sure the rinse water falls into the barrel so that the inside of the barrel may be washed and the residue all run down into the hole in the ground. We remove the barrel and fill up the hole with earth to keep other bees from becoming contaminated. We then place the washed frames in our refrigerator or deep freeze for twenty-four hours, then place them back into the scorched hive body, and hive a new swarm of bees. As a rule the new swarm will prosper.

The reason we scorch out the hive and wash the frames is that if the old colony died out due to poisoning, the heat and flames burn away the poison left on these wooden parts by the feet of the bees as they first entered the hive. Bees seldom seem to carry poison onto the honey combs, depositing it in such a way that we cannot wash them clean.

If they have done so and the colony dies out again in a few months, we burn up the entire hive with all its contents. It is far more profitable to buy or build a completely new hive than to waste more time hoping that another swarm will succeed where the others failed.

I have referred to yellow jackets, their inroads upon some hives and our remedy, and touched upon the subject of skunks. Our last great killer of bees in this area is a mysterious one—the north wind. I shall give the facts as my father and I have found them but I shall not attempt any explanation as I have none. Here is what sometimes happens.

Every year during late winter and early spring, we have winds from the south, then days of calm, followed by three days of wind from the north. This is ideal bee weather and our bees build up rapidly in numbers and gather large quantities of nectar and pollen. However, some years we have four or more days of north wind. In the morning after the fourth day of north wind we find several dozen dead bees on the landing boards of some hives. If the north wind continues, on the morning of the fifth day dead bees will be so numerous as to almost clog some hive entrances. Our strongest and best hives are always affected the most adversely. The longer the north wind blows, the more bees die. A sudden change in the direction of the wind and the dying stops instantly.

One year we had twenty-one days of north wind. This proved to be by far the most disastrous year we have ever known. Day after day we watched as thousands and thousands of our beloved bees succumbed to the deadly north wind. Every morning I took a stick and helped clear the masses of dead bees from each hive entrance. Finally the weather changed. The dying ceased. Our five strongest hives, the ones with a chance to set a new world's record, were a total loss. All the bees in them were dead and all the eggs, larvae and brood were dead in their cells for lack of nurse bees to keep them warm. Our remaining hives were weak in numbers but did have enough bees left to nurse and save their brood. Since each hive had a large number of brood, these hives built up again rapidly and were still able to make us a goodly amount of honey.

The above facts I have given in order to help allay the fears of new beekeepers when they see their bees begin to die if the north wind continues to blow a few days longer than usual. Fortunately for us as beekeepers, years may pass before this capricious north wind again plays havoc with our bees. When it does, we always hope for a change in the weather as that is the only thing that helps.

9

Moving Bees From One Location To To Another

LONG-TIME BEEKEEPERS who have come to visit us through the years have many times told me that bees are color-blind and it is useless to think we can use color as an aid in moving bees short distances from one location to another. I know that more recent experiments have shown that bees can distinguish at least some color variations, but these old beekeepers refused to believe the latest reports and I had no way to prove them in error. Bees are so intelligent and adaptable to circumstances in every other way that I could hardly believe they would be color-blind, especially since flowers bloom with such a riot of color. Years went by and I had no positive way to find the answer from personal experience. But one day on coming home with a load of long lumber requiring a red flag at the end, we noticed that the bees were returning to their hives so heavily laden and weary from their flight that many of them were narrowly missing the entrance boards. These two hives were on an unusually high stand because they were among our sheep, and sheep delight to tip over anything that can be toppled. When the bees missed their landing board, they fell to the ground and were lost in the grass until they were rested, after which they flew up to the hive entrance and went in. This seemed like a great loss of time and energy. How could we help our bees?

My father suggested tacking a piece of cloth to the landing board of each hive and letting it hang down. If a bee missed the landing, she could catch herself on the cloth and at least not go down all the way to the ground under the hive. No sooner said than done. I grabbed the red flag from our load and tacked it to our No. 2 hive which was having trouble. Then I found a piece of white cloth about the same size and shape as the red one and tacked it to our No. 3 hive next door to No. 2. The idea worked surprisingly well.

After a month the cloths became soiled so I took them off and washed them. When they were dry, I tacked them back onto the hives again. And then a most astonishing thing happened. Whereas hives No. 2 and No. 3 had always lived next door to each other in harmony, now the bees fought like crazy! Soon the ground in front of both hives was littered with fighting, dead and dying bees. I called my father to come quickly and help solve the mystery. For over an hour we watched helplessly.

Then my father said, "Ormond, are you sure you nailed the red and white cloths back where they belong? Didn't we have the red cloth hanging below our No. 2 hive? Now you have it below No. 3." Sure enough, I had unwittingly transposed the red and white landing cloths. In a minute I had them off and on again as originally placed a month earlier. The warfare ceased! All was soon as peaceful as before except for the many dead and dying bees scattered all over the yard.

The conclusion we came to was that the field bees, when returning loaded, were sighting in on their home hive by color. When I changed color on them they arrived at the wrong address, were met by the hive guards at the entrance, and the war was on. I was surely sorry to have been the thoughtless cause of so much distress among our friends the bees.

Having now established beyond any doubt that bees have definite color perception, my father began to consider how we might use this faculty in moving bees from one location to another. Honeybees have an astonishing sense of location both as to direction and height. When we bring home a swarm newly hived, or hive one of our own on our own place, we must immediately place it in the location where we want it permanently. A hive cannot be brought home, set down somewhere, and two weeks later merely picked up and moved over to a stand built for it twenty or more feet away. The hive box itself with the queen, hive bees, and brood, can be moved, but the field bees will go straight back to where the hive had originally been placed. There they will cluster, feeling completely lost, and there they will stay and die of exposure unless we move the hive body back again.

It is important to have the permanent stand completed beforehand so the bees may be placed where they are to stay. The next best thing is to set them on a box at the location and height they will be when the stand is built. After dark, or on a cold or rainy day when they are all in the hive, we can set them aside and build the stand. Then place them on it and they will not know the difference.

An alternative method would be to load up the hive and move it to a friend's place at least three miles away and leave it there for a minimum of two weeks—preferably longer—then move it back home to the new location. By that time the bees have forgotten where they were before and so accept their new surroundings.

If it is imperative that the hive be moved more than twenty feet and no friend is willing or able to take the bees for a couple of weeks, as a last resort we have had some success with the following method. We move the hive to its new location after dark in the evening, then take a hammer or larger stick and strike the hive all around so that we make some noise and jar the hive. The bees soon notice that a change of some kind is in progress. We do this until the bees are thoroughly aroused. After waiting an hour until the bees are quiet again we then scatter several handfuls of grass or straw all along in front of the entrance. When the bees come out in the morning, the grass partially blocking the entrance helps remind them that a change has been made and many of them will fly out in circles and relocate themselves. However, some will go back to the old location in spite of all that we can do. For them, we set a shoe box with an entrance cut into one end, on the old stand so the bees returning there may go in and be sheltered. Then in the evening, or sooner if need be, we empty the shoe box of bees in front of their hive. Most will go in but some always fly back and are lost. If some of my readers have found a better way I would be glad to hear how it can be done.

If a hive of bees must be moved just a short distance, say up to twenty feet, it can be done easily if one has plenty of time and patience. Each day we move the hive sideways, forward, backward, up or down, depending on the direction wanted, about half the width of the hive or eight inches. The bees notice the difference in the location of the entrance and buzz around awhile but then adjust to their moved hive and carry on as usual. Eight inches a day, moving every day, is as fast as we have been able to go safely.

Once we wanted to move a hive a full twenty feet. We started out as described above but after a week (four and one-half feet on the way) my father decided to move them along a little faster. Maybe they would stand moving at the rate of sixteen inches a day. Well they did for a few days but we could see it was with obvious reluctance. However, Father kept moving them along and had reached to within four feet of where he wanted them when all the fielders suddenly flew back to the old stand. There was nothing for us to do but move the entire hive back to where it had been before and start all over again. We did, and in due time with a little more patience, we got them moved to where we wanted them.

This past summer my father began experimenting with a new and faster way of moving bees short distances sideways, using the bees' color perception ability. He lay a white one-by-twelve inch board ten feet long on the ground pointing away from the hive. He let this white board remain there for three days while the bees became accustomed to flying over it when returning loaded to their hive. Then he moved the hive sixteen inches and the board also sixteen inches so as to keep it in front of and still pointing straight away from the hive. The returning bees flew directly to their hive entrance without hesitating as they normally would have done. Apparently they used the white board as a guide to find their way home. A few times he moved the hive as much as two feet a day and still the bees kept with him until he had moved them to their new location.

Caution: The white board must be left in front of the hive in its new location for several weeks before it can be picked up and removed. Once my father took the board away as soon as he had the hive moved and the field bees flew back to the old stand, compelling him to move the hive back and start all over again.

Nevertheless the white board method shows promise. Whether it will work with every hive we do not know. It takes many more than one or two successes to set a hard and fast rule in beekeeping. One hive differs as much from the hive placed next to it as one family differs from another family among human beings. In fact, I find bees are very much like human beings in many ways and that is why I often think of them as my beloved "buzzin' cousins."

10

Early Spring Build-up & First Supers

HOW EAGERLY we and our bees await the coming of spring! We all look upward for the first wild blossoms. Look upward someone might say? Yes indeed for here in Santa Cruz County the first wild blossoms to produce both nectar and pollen in quantity are those on the eucalyptus trees. Some of these trees grow to be really huge—four feet in diameter at the base and almost 200 feet tall. They are great spreading trees with innumerable pale yellow blossoms two inches across.

These trees are remarkable in that they begin blooming about January first of each year and bloom until July first. They often do not bloom all at once but each tree blooms one branch at a time. The trees are so tall that we must look very closely on January first to find any blossoms. But a first little branch with its hundreds of blossoms is up there somewhere because the bees go to and from the tree. On looking closely we too can find the branch the bees have found. Once started, these trees continue to bloom in spite of light frost, snowflakes, or hail. The bees work on them every sunny day to excellent advantage. We rarely get any pure eucalyptus honey because the bees use it themselves for brood rearing. It has an excellent flavor.

About the last of February the first of the three species of wild blackberries common to our locality begins to bloom. The combination of eucalyptus and wild blackberry nectars makes a delightfully delicious honey. We could never produce enough to supply the demand. I might suggest that because of obvious variation in types of vegetation from one locality to another the reader become familiar with the blooming schedule of local wild flowers.

After January first we see a noticeable change in the attitude of the bees. As soon as the morning sun warms the front of the hive the field bees are up and away. Soon they come back, some with pollen and some with nectar. Usually in this early part of the season about half of the bees return home with pollen and the other half with nectar. However, this is not always the case. The hive bees seem to be able to communicate with the field bees and tell them which is most needed. Sometimes we have seen all the field bees of a particular hive returning with huge loads of pollen for an entire day. None of them were carrying nectar. But the next day they were on a half and half basis again. Other times, we have seen them carry nectar only until the apparent shortage was corrected and then they carried pollen again also. This is just one of the reasons why I believe that honeybees have an intelligence above the level of instinct.

If our weather is mostly clear during January and February the bees build up rapidly in numbers. From a low of 15,000 to 20,000 in December, each hive as of March tenth may have up to 40,000 bees. From this date they build up even more rapidly. By April eighteenth, our best hives, the ones with a chance to set a new world's production record, have about 115,000 bees per hive. By this time we have given each hive six or seven supers for the bees to live and work in as well as space to store excess honey.

For those who may one day want to try for a new world's honey production record, or just learn more about their bees, let me say right here that we can learn a great deal about each hive of bees by getting up at night and listening to what the bees are doing. I press my ear against the side of each hive body and listen, then listen to each super in turn until I get to the top super. Then I go to the next hive and listen in the same way, listening to each hive we own. In this way I am able to tell by the voices, noises, rustlings and fanning roar inside the hives what the bees are doing at present and what they are likely to do in the near future. I am also able to tell a few days in advance which hive is going to swarm by the "squeaks" the young queens make in their cells while they are still sealed in. They give these sounds most often just at or after dusk.

Bees have a language which, when mastered, is most helpful to any beekeeper whether he is in search of a new world's production record or just wants honey for home use. Having had bees all my life I suppose I learned it naturally and had no idea how hard it would be to teach someone else until some years ago when I tried to teach it to Gary my neighbor. For several years he had often worked closely with me, had acquired two hives of his own and become intensely interested in bees. Under my supervision he had remarkably good success with his hives. He came over to see me almost every day, or I went over to see him and observe his hives. He had a great desire to learn the bees' language. For more than three months we listened to our hives both day and night and I explained as best I could what the various sounds in the hive meant. At the end of the season he was sure he had mastered the language.

"Next year, Ormond, don't help me with my bees," he said. "Let me work them on my own—and I'll race you for a new world's record."

"Fine!" I agreed. "I'll be glad for the competition."

That next season I saw little of Gary, but when I did his face was grim and often puzzled. At the end of the season he reported that one of his hives had swarmed several times and made no excess honey at all. The other had done well, I thought, for he had extracted 125 pounds of delicious honey. But Gary was disappointed.

"I didn't learn the language," he told me at season's end. "Will you take over the care of my hives next year? I'll be away at college."

"Be glad to," I answered.

Gary had done well, everything considered, for it takes time to learn the art of beekeeping. He had good bees, for next season each hive again produced more than 100 pounds of honey.

I sometimes get up at eleven o'clock at night and listen to all our hives. The next night I go out at one A.M. and listen, the following night at three A.M., and the last night at five A.M. If on or about April twentieth we have a hive that is still roaring inside, with fanners out on the entrance board at five o'clock in the morning, that hive has a good chance to set a new record. At the very least, we will harvest a great amount of honey from that hive provided, of course, that we keep alert and put on and take off supers as indicated by the message from within.

This listening at night also helps tell us when to take off a super. For instance, if on the night of April tenth, at one a.m. there is a great roar going on in a top super, it means the bees are hard at work in that super. When I listen again on the night of April eighteenth, and the roar has almost ceased, it indicates that the super is full and sealed, ready to be taken off on the next warm day. When a super is finished, the bees almost all forsake it and go to work where they are most needed. However, even though we know that the top super is full and sealed we do not remove it from the hive and extract the honey (unless we are desperately short of supers) until we have added at least a total of five or six supers under it and they are well along toward being filled also. This is to encourage the bees to fill the lower supers we have added. Bees, like people, feel more secure and content, ambitious and active when they have a substantial golden store tucked away in case of need. Where the weather is cool it is of great importance to know the bees' language so that we do not open a hive unless there is urgent need to add or take off a super.

"How do you know when to put on the first super after the early spring build-up is well under way?" I am often asked. This is a difficult question to answer because of the many and varied factors involved. Let me discuss the difficulties at this point and give our answer a little later. For instance, if the weather during January and February is relatively warm and sunny, supers will need to be placed on the strongest hives the first warm afternoon that comes along after February twentieth *unless* we are going to have a late winter with frost in the early days of March. In that case we wait until about March fifth to minimize the danger of severe frost so as not to have unnecessary cold air space in the hive. The dates given above are for our sea-level Santa Cruz area. For bees located at higher elevations or inland regions with cold winters the dates would have to be adjusted to correspond to the climate. Weather reports and forecasts are important to us as beekeepers.

Remember, bees maintain a hive temperature of between ninety and one hundred degrees Fahrenheit to preserve the lives of their eggs and baby bees. If we put on a super and then have several days of cooler weather, the bees may not be able to maintain the necessary hive temperature and some of the young brood may die. In this event we will soon see hive bees dragging out four to eight dead white larvae, possibly for several mornings in a row. Dead white larvae on the landing board at other times in spring are usually the result of inroads of wax moth worms cutting through some of the brood combs. This is a problem the bees can handle themselves. Our efforts to aid them are more of a hindrance than a help.

Then again the queen, who has been laying thousands of eggs a day, may suddenly feel the cold and stop laying for a few days or even two weeks. In either case, as beekeepers, we sustain a substantial loss as our cold snaps never last long and we want the bees to continue building up rapidly in order to gather as much of the now abundant nectar as possible.

Again, if the hive needs a super and we do not add one, the bees become overcrowded and start so-called swarm cells, really queen cells. These are oversized cells the bees build so the queen can lay an egg that will rear a new queen. In two or three weeks' time the old queen flies away with approximately half the bees in the hive, a new swarm, to take up housekeeping elsewhere. Hopefully, the queen and her swarm will cluster on our own property so that we can hive them. Therefore, it is of the utmost importance to give them more room—a super—before they build queen cells. A hive that swarms will not make nearly as much honey as one that does not swarm. Our goal is to keep our bees from swarming not more often than once every other year. To minimize the bees' early swarming instinct we have of late years left a medium depth super on each hive over winter. Our hive at this time is made up of parts in this order: A bottom board, the hive body, a queen excluder, the super (well filled with honey last August fifteenth but now partially emptied), the top cover, an insulating cover, a roof, and several bricks on top for weight. This arrangement has an added advantage. When putting on the first super of the season, we raise the super already on the hive and slip the new super beneath it but above the queen excluder. The bees and honey in the old upper super will encourage the bees to begin work immediately in the newly added super.

Before placement of a new super it should be set out in the sun for up to half an hour so that it can warm up from a storage temperature of approximately fifty degrees to the hive temperature of ninety-five degrees. This solar preheating of the super saves much labor for the bees as well as honey that they had stored for their own use or for us.

Caution: All winter long or early in the season or whenever the hives are put out to begin the spring build-up, always remember to place the weights on the roof of the hives. These weights keep the covers pressed down to the top of the hives and conserve heat otherwise lost through cracks. We use concrete blocks weighing from twenty to forty pounds on each hive until the summer temperature reaches eighty degrees or above.

But the question still remains, "How do you know when to add the first super? Is there a positive answer?" Yes, at least in part. I am taking it for granted that as beekeepers we are taking time to observe each hive we own every warm afternoon. This is an absolute necessity if one hopes to reap the maximum amount of honey possible per hive. Standing by a hive on a warm afternoon we count the bees *returning loaded* to the hive. One cannot hope to count the bees leaving the hive because they take wing too rapidly. But count the returning bees. When the bees return at a rate of 100 to 120 per minute and the temperature is seventy degrees or slightly above, we put on a super. As an aid in counting we divide the hive entrance into two parts by placing a white stick on the landing board as a divider. My father stands on one side of the hive and counts; I take the other side. Together we can obtain a more accurate total. One often has to take his chances with the weather and other factors when putting on this first super. We have never found any two locations exactly alike; there is endless variation. A few years' experience in one's own area provides the best instruction. It takes time to acquire expertise in beekeeping as in every other endeavor.

Let us say that we have had a normal winter. The bees have built up strongly since January first, and it is now February twentieth to March tenth and time to put on the first super of the season. A beginning beekeeper who has no previously drawn super combs should add a super with starter sheets when the bee count reaches approximately 115 bees per minute or possibly more if the weather is still quite cool. This first super always entails a calculated risk and there is no easy answer.

A second-year beekeeper, one who has successfully brought his hive through the first winter with a queen excluder and super above, should now raise the super that has been on all winter and slip this second super under it but above the queen excluder as stated earlier. If this added super has wax combs drawn the year before, the bees may well need a third super within twenty days. But if it has just wax super starter foundation in each frame, the bees must first make wax to draw out the foundation to make room to store honey. This usually takes about twenty-four days or four days longer than with previously drawn combs. The above figures are for an average hive. A weak hive would take longer and a very strong hive fewer days. There is no substitute for experience but one will soon learn how to evaluate his hives.

In this early part of the season we rarely give a hive a second super with all drawn combs. Rather, we take four drawn combs, placing two on each side of the super to be added, and place six frames with starter sheets, between them. The drawn frames on each side are a great encouragement to the bees to draw out the starter sheets in between. Bees can draw out these sheets in a few day's time if they can be encouraged to put their minds to it.

Now, as the weather warms up, we begin again to open the hive entrances which we had closed about three-quarters shut with entrance cleats last fall and winter. We use, in the height of the honey flow, a wide open entrance which is three-quarters inch high by the full width of the hive. By the time we put on the second super the bees should be crowding the ten inch long by five-eighths inch high opening we have had up to now. So we make the opening a little longer for easier access for the bees and also to allow them more fanning room for evaporating the nectar and ventilating the hive. We also enlarge the entrance during the heat of the day and contract it again at night if this seems desirable to keep the bees from crowding the entrance too badly on a warm day when the nights are still cold.

Add more supers at eleven to twenty day intervals depending upon the strength of each hive. We put them on in the following order for best results: Slip the first super of the season, we will call it No. 2 super, under No. 1 super which was left on all winter. Then slip No. 3 super between No. 1 and No. 2. Slip No. 4 super under No. 3 but above No. 2. Slip No. 5 super away down at the bottom just over the queen excluder. Slip No. 6 super above No. 2. Slip No. 7 super just below No. 2 and above No. 5.

Seven medium depth supers are all we use for one hive. When yet more room is needed we take off super No. 1 and extract it replacing it about three supers down for the bees to refill. Sometimes we get as many as twelve well-filled supers from one hive in one season.

Here is the reason for our method of adding supers as stated above. Super No. 1 has been on all winter and still has some honey in it and a goodly supply of pollen. The bees need the pollen, and this super must be raised only because the bees must be encouraged to

Addition of Supers

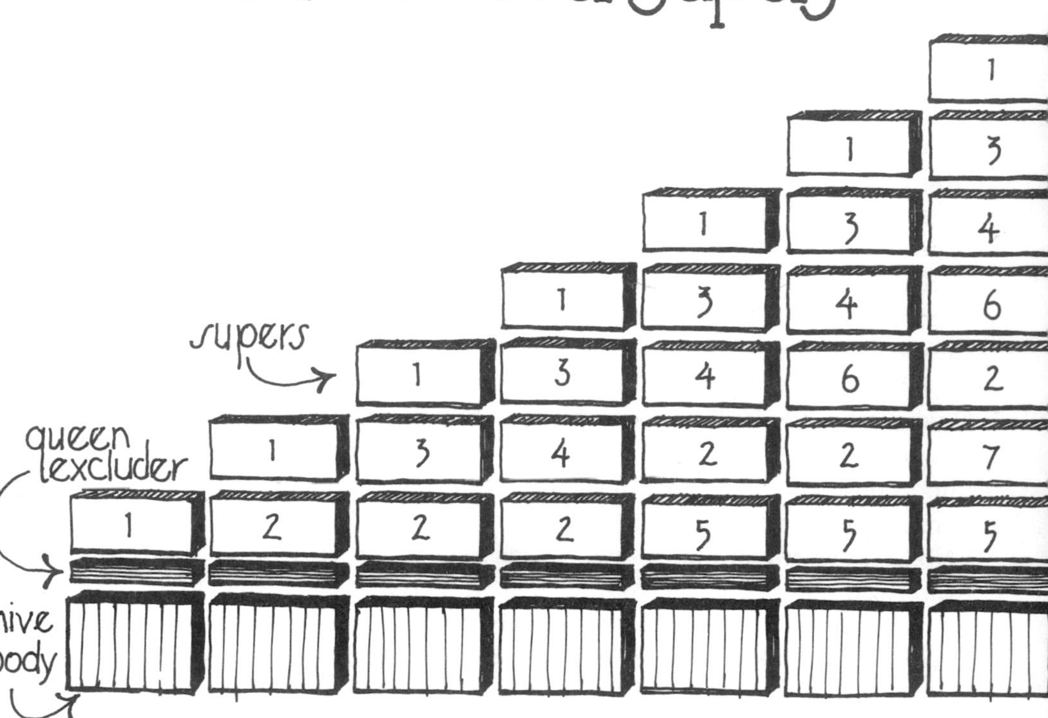

fill super No. 2. The bees will store much pollen in No. 2 since it is now nearest the brood chamber so leave No. 2 on the bottom as long as possible to make it easy for the bees to use their pollen as required by their brood. However, by the time we need No. 5 super the hive is very strong—upwards of 100,000 bees—and the entrance is wide open. The hive body (brood chamber) is full of bees. Super No. 2 is filled to capacity with honey and pollen. Filled to capacity, in this one instance, does not mean that every cell in the combs is full and sealed over. It does mean that the bees have filled and sealed all of the cells that they will do willingly before they think of swarming. They almost invariably leave a rather large half-moon-shaped area on each frame with partly filled open cells to use as work, storage, and living space. The higher supers are becoming well filled also.

Something must be done to make more room just above the brood chamber or the bees will most certainly build queen cells and swarm. We want to avoid swarming this year at all costs (this hive swarmed last year) yet continue to encourage the colony to build up even more—up to 125,000 bees or an even larger number. This hive has a chance to set a new world's record! We raise all four supers already on and slip empty No. 5 with fully-drawn combs in at the bottom to relieve the need for additional work space. Soon No. 5 will have much pollen so we leave it there for the remainder of the season and add empty supers just above it as more space is required. Often we leave this No. 5 super on for the winter provided it is well filled with honey and pollen. If a few frames seem to be not so well filled we take them out and replace them with well-filled frames from No. 1 super, which by season's end will have been filled and extracted two or three times. Since an extractor removes the honey but never the pollen, we choose for winter feeding those frames with a maximum of pollen as well as honey. We leave a well-filled super of honey on each hive for the bees' winter food. *Never rob your bees*—only take from them the excess honey that they do not need. Always at all times of the year, leave for them plenty of honey so that every day they have a feeling of well-being and abundance.

The foregoing illustrates in part what I mean by "encourage" your bees to work for you. One elderly man who has never had bees of his own but knows something about them, often accuses me of running a "honeybee sweat shop." He has had friends who have kept bees and they never got more than one or two supers of honey from each hive. Since my father and I have obtained so much more honey, he insists we must be using some ungodly method of "beating our bees." This is not true for we love our bees. When working in our apiary we always talk to our bees and give them our love. They soon get to know their keeper's voice and seem glad to have him come around. This is especially true if he treats them to a little honey when there is no honey flow.

We cannot possibly *force* bees to make honey for us, for they are the most independent little creatures the Good Lord ever put on this earth—but we can do much to *encourage* them to work for us.

11

By the time spring, according to the calendar March twenty-first, has come, our bees have increased greatly in strength and are making honey at a rapid rate. At this time we check each hive to see that it has an extended landing board. If the hive has a bottom board extending two or three inches beyond the entrance, now is the time to add an extension board that will lengthen the landing area to nine inches long by the full width of the hive. We need to be sure the extension board is on a level with the hive entrance, neither above nor below it. Once such a board is installed the bees begin to use it. The heavily laden homecoming bees that before would have missed the landing, now are able to alight on the extended board and can crawl on the level to the entrance. This extended landing board is a great improvement, where practicable, over our red or white vertical landing cloths tacked to a short entrance.

Soon the "fanners" will be seen taking their places far out on the landing board. These are bees that anchor themselves to one place and then fly—that is they vibrate their wings as though they were flying—but their bodies remain stationary. They do this to keep air circulating throughout the hive. The bees seen standing in rows fanning on the landing board are the last of the thousands of bees throughout a large hive that are fanning with their wings to keep fresh air circulating in all parts of their home. On a quiet day one can feel this movement of air coming from a large hive while standing four feet away. Early in the season this outgoing air occasionally has a sour green fruit juice scent due to the evaporation of certain newly gathered nectars. Later, as more supers have been added and the nectar has begun to ripen into honey the scent becomes sweeter. When much nectar has been converted into honey and the filled cells have been partially sealed over by the bees, a truly pleasing sweet scent comes from the hive. Then we know that we can begin to take off supers of honey very soon.

Caution: When opening a hive, either to slip in another super or later on to take one off, always work at top speed. *Try to have the hive open not more than four minutes.* Speed in this part of the operation is absolutely essential to the production of honey here in Santa Cruz and doubtless in many other places in the world.

The reason is that though there are many sunny days in the late winter and spring, yet we have few really warm days. We never open a hive unless the outside temperature is seventy degrees or above. Even then we work at top speed. Before opening a hive we have everything ready in advance: Smoker lighted and going well, extra pieces of burlap sack in a pocket in case additional smoker fuel is needed, hive tools in hand or pocket, bee brush, gloves on (beginners), veil on (beginners and sometimes myself), extra super with frames all in place, and an extra super box without frames but with a cardboard or plywood top and bottom to keep the removed filled frames of honey clean and also free of investigating bees. Think out in advance what will have to be done and then do it without any false motion or loss of time. It is better to have everything laid out and not needed than to need something and not have it handy.

Let us say the outside temperature is seventy-one degrees and the normal temperature in the hive is ninety-five degrees. The moment the hive cover is pried loose and removed, there will be a noticeable "whoosh" of hot air in one's face. The warm air of the hive has gone out with a rush because of the lower entrance opening allowing cold air to come in. The hive continues to cool off rapidly all the time it remains open or apart, so we examine the frames in the supers as rapidly as possible, then take off or add frames or supers and quickly reassemble the hive. I cannot stress this fact too strongly. Skillful handling of the bee brush speeds up the work and saves the lives of many of our hard-working little servants. The life of every bee is precious in the eyes of a true beekeeper.

"Where do bees get the heat to rewarm their hive after it has cooled off?" I am often asked. The answer—the bees eat more honey for a time, and since it is such a rich food their bodies throw off more heat. But this rewarming of the hive costs us honey for the bees are compelled to gorge on some of the honey they had stored as excess for us. Honey is money so as beekeepers we need to do everything possible to assist our bees to increase their stores rather than to deplete them.

Let me illustrate. An older man of my acquaintance had owned two hives of bees for several years but he never got any honey from them. Then he heard of me and asked me to come over and look at his bees to see what was wrong with them. One quite cool morning about April fifth, I stopped by to see him. As soon as we came to his hives, he lifted the cover off the first one. I was so astonished that I shouted, "Put that cover back on quick!" But he would not do it. I explained as quickly as I could that he was practically freezing his bees. Still he would not replace the cover as he said his bees were used to being uncovered for he did it every morning to see what, if anything, they had done the day before. Poor bees! It was not their fault that they could produce no honey. This type of meddling is a common and, for the bees, most distressing situation perpetrated upon them by many otherwise kindhearted beekeepers.

Some beekeepers cannot seem to contain their curiosity, they have an insatiable urge to see what is going on inside the hive each day. Others, when opening a hive in March or early April, find that the inside of the hive cover is wet with water and they think they can help the bees by taking off the cover every morning and wiping it dry. This is sheer folly. True, the wetness some years may be so bad that the combs of wax become covered with mold, but let the bees alone and as they increase in numbers and need the room they will polish up these combs so neatly that one would never have known that they had been moldy. A beekeeper can help to prevent this wetness for it is rarely found when a hive faces east and the front of the hive as well as the entrance receives the heat of the early morning sun. Then the bees can circulate warm dry air through the hive and take care of the wetness problem themselves and at the same time retain the necessary hive temperature for maximum honey production.

There is another type of wetness that may be observed as soon as the bees find a substantial source of nectar and begin the evaporation process to convert the nectar into honey. Nectar has an average ratio of eight to one—eight parts water to one part honey. When the bees have brought nine pounds of nectar to their hive they will have stored, in about six weeks in our damp area, but one pound of finished honey. Where the air is hot and dry the conversion process is completed much more rapidly. What a massive evaporation and conversion problem the bees must wrestle with! Were it not for the fact that they are such tireless workers they would most surely give up in despair. For a hive to make eighty-five quarts of excess honey for us, as they have occasionally done, they had to visit enough wild flowers to gather and carry home to their hive 765 quarts or 2,295 pounds of nectar—and all this in a period of a few months.

At the height of the honey flow we often find a situation where a beekeeper can render needed assistance to his most productive hives. During this time the bees bring home a great deal more nectar than they can evaporate in the daytime, so they run a night shift—sometimes all night long if they are unable to get the job done sooner. In this event when we go out to look at the bees about seven o'clock in the morning, we will find a puddle of water in front of the hive entrance on the landing board. This has been caused by the warm moist air of the hive coming out and meeting the colder outside air, causing condensation. We are always delighted to see a pool of water, for it means that we have a strong hive hard at work. However, we soon observe that many of the early morning fielders leaving the hive become trapped in this water and have difficulty in getting out. To assist the bees, we gently wipe away the water with an absorbent cloth or sponge. If we stand well to one side of the hive and do not obstruct the flyway, this can be done without veil or gloves.

In many areas of the world bees must be provided with water during the heat of summer, and shade also. We have kept bees in dry inland locations where our bees posed a real problem for our livestock. On hot dry days the bees frequented the watering troughs of the farm animals in such numbers that the sheep, cattle, or horses were in danger of being stung when they went for a drink of water—not because the bees were hostile to the livestock but because great numbers of bees were crawling all over the damp inner surfaces of the troughs—some bees even getting into the water where they were in danger of being drowned or swallowed. If an animal inadvertently rubbed its chin against the side of the trough it was almost certain to be stung because it unknowingly crushed a number of bees, and such frightened bees will sting friend or foe. I know of no completely satisfactory bee watering device that will keep all bees away from livestock watering troughs. However, almost any available widemouthed container provided with a wooden float, filled with water, and placed near the bees, will help.

In our Santa Cruz area close to Monterey Bay our bees never need supplemental water and seldom shade. On rare occasions on a hot day we see masses of bees on the outside of the hive hanging in a cluster from the landing board. In that event we immediately lean a sheet of corrugated aluminum roofing against the hive to provide shade. Of course in some regions permanent shade must be provided.

Honeybees need a little salt to keep them working at maximum efficiency. We first noticed their need for salt when we gave unrefined table salt to our sheep. We found our bees picking up salt crystals that the sheep licked off and dropped from their lips after a sheep had wandered away from our freshly filled salt feeder. That gave me the idea to take a salt shaker and give each hive a helping of salt every few weeks during the honey flow. One or two shakes on each landing board at midmorning is sufficient. If a hive lacks salt, a field bee, crawling out of the hive on its way to take wing for a load of nectar or pollen, will suddenly stop when it comes to a salt crystal, lick it, and then grasp the crystal and carry it into the hive. Soon other bees come out on the landing board and gather up as many of the remaining salt crystals as they need.

Drones are the large male bees that inhabit the hive. They have no stingers so we need not be the least bit afraid of them, though the first ones out in the spring make such a loud buzzing as they fly about that they sometimes startle me. They are really quite delightful fellows. After the early spring build-up is well under way, the queen bee will begin laying drone eggs in large especially built cells.

In twenty-four days these big fellows are grown and hatched, and soon one or two may be seen walking around the entrance of the hive.

My father and I always look for the coming of the drones with real expectation. Their presence again in a hive is cause for rejoicing, both for the bees and for us. Their primary purpose in life is to fertilize the new queen when the bees swarm. It takes from three to five drones to do this.

"Then of what use," one may ask, "are the hundreds of other drones found in every strong hive during late winter, spring, and summer?"

These drones play a vital secondary role. They are big hot fellows and they help to keep the hive warm. Here, where our weather is cool and comfortable for us as human beings, it is almost cold for our friends the bees. We rejoice when we see drones in every hive, for then the danger of losing brood during a sudden cold snap is greatly reduced. True, drones do eat a lot of honey and do no work such as gathering nectar, making wax, feeding the baby bees, or keeping the hive clean. They do also sometimes run all through the hive and tease the girls and appear to make a general nuisance of themselves. However, they seem to be an encouragement to the working girls in spite of their antics. It is certain that in our area the hive that makes seventy-five or eighty-five quarts of honey in a season always has a large number of drones. I would no more think of killing a drone bee than I would of killing a worker bee. We need every one of both kinds.

We prepare a new hive with full sheets of wire- or plastic-reinforced foundation comb and then let the bees do what comes naturally—and naturally they will cut out some of this comb and build large cell drone comb in its place. So be it. They are wise little creatures; let them go about their business and wish them Godspeed.

Drones are privileged characters in that they can visit any hive in the apiary they please and the hive guards will admit them and let them come and go freely. They seem to be able to tell which hives have ripe queen cells with young queens about to emerge.

One warm afternoon I watched several drones buzzing around and alighting upon the extended landing board of our No. 11 hive. It was early in the season and this hive still had a restricted entrance cleat. Most of the drones entered the hive and staved inside for at least a few minutes before coming out again and so caused the guard girls no trouble. Drones mate with the queen while flying around outside, so the drones do not enter the hive for the purpose of mating at this time.

One handsome strapping young drone seemed unable to make up his mind just what he wanted to do at one of our hives. He ran in and out of the entrance again and again, pulled or bumped the guard bees, and in general made himself a downright nuisance. Suddenly, without warning, three of the guard girls grabbed him, one by the head, one by the tail, and the third wherever she could get a good hold. Together they yanked him along toward the end of the landing board. For several minutes it was a regular tug-of-war. When he could get a good footing, he would lunge back toward the entrance taking his bouncers with him. Then they would roll him over, giving them the advantage again, and be on their way once more. He struggled wildly but they continued to roll, push and pull until they got him to the end of the landing board where they unceremoniously dumped him over the edge onto the hard ground below! They had cooled his ardor, believe me. I watched as he carefully checked himself all over to see if he still had all his legs and other members. Finding that he had, he slowly crawled out from among the clods into an open space in the sun, rested a short time, then took wing and flew up to the landing board again. This time he very cautiously approached the entrance, was checked over by the guard bees, then admitted without difficulty. That was the last I saw of him. Could it be that honeybees have mastered the art of "women's lib?"

In fall I feel sorry for the drones because at that time of the year the worker bees drive all the drones out of the hives. The queen is laying few if any eggs, so there is no need for additional heat. Stores of honey must be conserved for the good of the hive. Since there is now no useful work for the drones to do, the worker bees ruthlessly force them all out to die of exposure. In a beehive the age-old, God-given law, "He who does not work shall not eat," is strictly enforced.

12

Keeping Bees Gentle

THERE ARE NUMEROUS devices we can employ to keep our bees gentle. In recent years we have kept five hives within fifty feet of our honey salesroom and our back door. Many people have parked their cars in our barnyard and walked right past the beehives, but only one person was stung as far as I can remember and that man got a bee in his hair. Why did ours not sting more often?

There are a few basic rules to remember when setting up hives and pedestrian walkways. Always arrange the hives so that people do not walk in front of the hive entrance or directly across the bees' flyway. Rather, have them walk past the side of the nearest hives.

Another very important thing is to keep something in more or less constant motion near the bees. Bees become accustomed to this motion and pay little or no attention to people passing by within a few feet of the hive. A plum tree growing beside the flyway, but close to a hive, provides excellent motion because the wind constantly shakes the branches. If there is no low tree near the bees and the walkway, one can drive two tall stakes into the ground about six feet apart forward of the hive but to one side of the bees' flyway. A wooden crosspiece wired to the top of the stakes provides a support from which to hang a piece of cloth or burlap sack about three feet square. The wind will keep this "wave cloth" flapping enough to help accustom bees to passing people or other activity.

Three years ago a woman bought two hives from me. We stood within a few feet of these hives and talked bees for an hour and no bee seemed to notice us. Three weeks later she came back and said the bees threatened to sting her if she even came in sight of them. She had placed her hives in the center of an open three-acre field. I asked her if she had put up any wave cloths near her hives and she said she had completely forgotten about them. In her case she needed three wave sheets, one on each side and one directly in front of the hives thirty feet away. She put up the cloths and has had little more trouble.

I have stated before that bees will not sting unless they are frightened or offended. In the foregoing illustration the woman frightened her bees when she came around them, for they were not used to strange motion. The following is an example of a man who offended his bees. This man also bought two hives from me some years ago. He was delighted with his bees after he had taken them home. For two weeks he sat on a box in the sun beside his hives and watched his bees work. Then came a cold frosty morning. He put on an old coat and a very worn old woolen cap, went out and sat down by his bees. Ouch! He was stung in a moment, and then stung again. His bees chased him all the way into his house. You guessed it—he came to see me.

"Those bees of yours," he said, "have gone completely crazy!"

I looked at him and could not help but smile. "Were you," I asked him, "by any chance wearing that old woolen cap you have on now?"

"Why yes, I was," he replied. "It was cold that morning too."

"All right," I said. "There is your answer. Do you remember I warned you of this? Sweaty woolen clothing of any kind offends bees!"

"That you did!" he said, "but I completely forgot." He went home and had no more trouble with his bees.

Bees do not like the scent of woolen clothing. This does not mean that they will always sting the wearer of such garments, but it does mean that the bees become nervous and are inclined toward using their stingers. Bees like sheep well enough. They are always friends. A bee will not sting a sheep no matter how much of a nuisance it makes of itself unless it tries to eat a live bee which has settled on a blade of grass to rest. Then the unfortunate sheep gets stung in the mouth and that hurts just as much as it would you or me. But woolen

Bees in Close Proximity to People

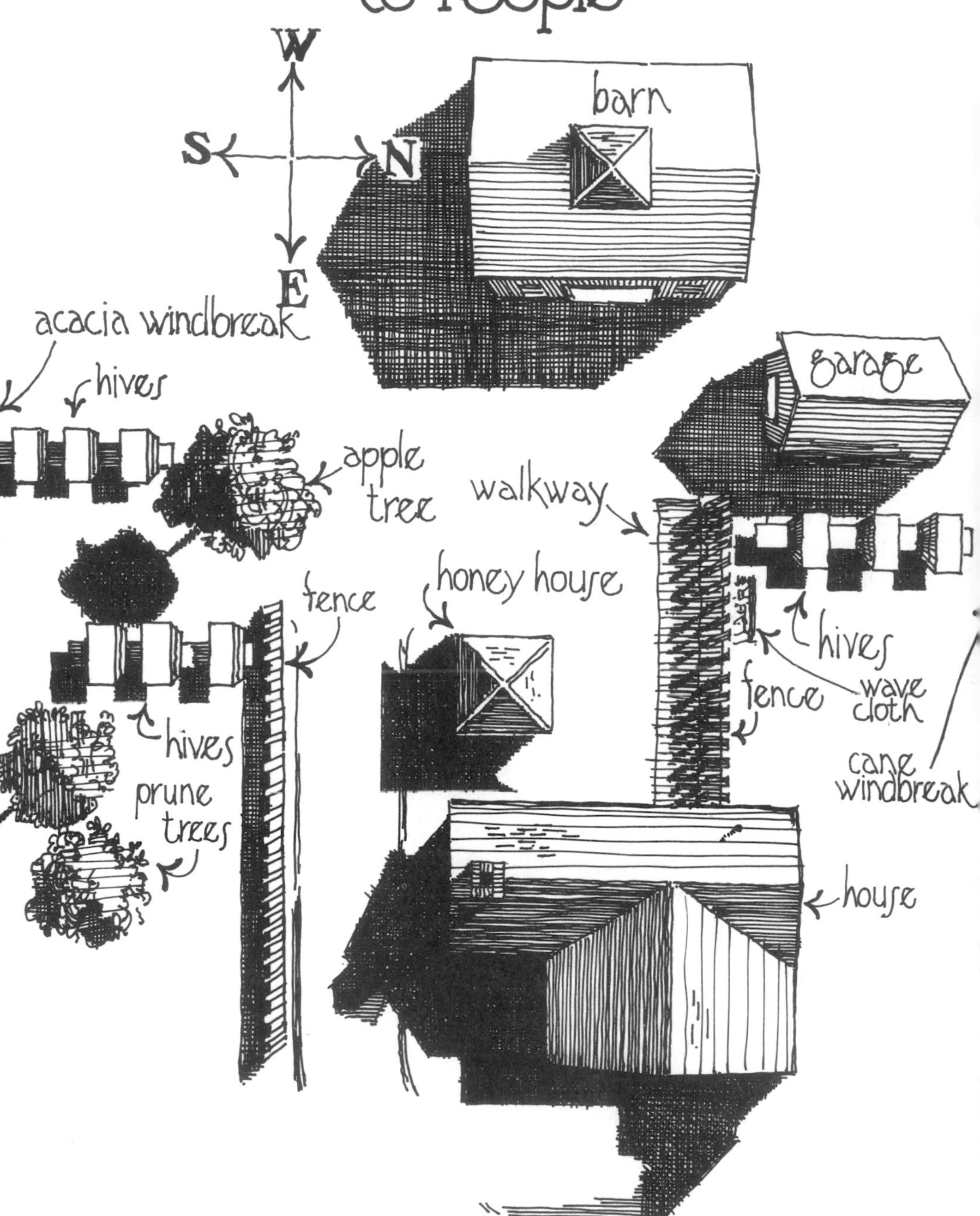

clothing and bees do not mix well. An old well-worn and sweaty woolen cap is particularly offensive—so is a sweaty scarf. I used such an old woolen scarf one warm spring day to test this statement to see if it was really fact or fancy. Wearing clothing all made of cotton except an old scarf, I stood directly in front of one of my gentlest hives. Ordinarily, these bees would fly all around me and make no effort to sting. I had done this many times before. But this morning, within a minute, a bee began to circle my head and give out that wing produced, high-pitched sound that always means, "Get away from here, get away from here or I'll sting you." Around and around she flew for more than three minutes. It took all the nerve I possessed to stand still that long. And then I got it, but good, in the back of the neck just above the scarf. Poor little bee, she did not want to sting me, but apparently it was her duty to guard the hive from offense and since I just would not take warning and leave, she had no alternative other than to sting. Some of our synthetic clothing is also offensive to bees but I have not yet been able to pinpoint which ones.

Once we received a frantic telephone call from a young woman who said that a great swarm of bees had come into her back yard about an hour earlier and had clustered on a wheel of an old trailer parked there. She had locked her four children in the house because these bees seemed to want to sting. Would we come over immediately and hive them? I told her that we would be right over. We dropped everything and were loading up our car when our neighbor Gary strolled down the driveway.

"Hey," he said, "you are going after a swarm of bees. May I go with you?"

"Come on," I answered, "but make it fast. Run home and get on a shirt and jacket. Bring your veil and gloves. We'll be by in a few minutes." He was ready when we passed his house so we picked him up and took off. On the way Gary showed me a brand new pair of gloves he had bought for just such an occasion. They were like any other pair of cloth work gloves except that these had a three-inch black cuff.

When we arrived at the house, we found things exactly as the woman had said. It was a huge swarm, located on the inside of the wheel and around the axle. I told Gary to don all his bee gear and my father and I would put on our veils, too, for this was going to be a difficult swarm to hive, as they were in an almost inaccessible location.

It was well for all of us that we took no chances. Father and I were getting our equipment and hive set up under the trailer when Gary came near to help. Almost instantly I heard the high-pitched whine of angry bees. A moment more and there were hundreds of angry bees and they seemed to be particularly irate at Gary. He felt inclined to run and I told him to run away, by all means! Just then a bee stung me on the hand. With so many more angry bees flying in every direction I began to wonder if I had not better follow Gary. However, I was lying on my stomach under the trailer at that particular moment so could not take off and run. When I was out from under and could flee, there was no longer any need—all the angry bees had pursued Gary. After they had chased him up the street, they came back but were no longer angry. I could not understand it. Father and I hived the swarm in the usual twenty minutes without further difficulty. We picked up Gary and started for home.

"Say," Gary said, "did you see those bees light into me?"

"I surely did!" I replied. "Did they sting you badly?"

"Well, yes and no," he said. "They didn't sting me anywhere so that it hurt, but I pulled twenty-one stingers out of the cuffs of my new gloves." Then we knew the answer to the riddle—the cuff material of his gloves had offended the bees.

Here let me say that honeybees have personalities of their own, much like human beings. Some hives are gentle, others are cross. This seems to be due for the most part to the temperament of the queen. As stated before, it is my belief that bees will not sting unless they are frightened or offended. However, as with people, some are more easily frightened than others and some are more easily offended. Just as we are sometimes unaware of how we may have offended someone, so we may be completely unaware of how we have offended our bees. Also, not all people or bees are offended by the same set of circumstances. First of all we should try to find out what it is we do that angers our bees, for more often than not it is the beekeeper who is at fault. When working with our bees we always have a smoker going and use a few large puffs of smoke at the entrance and under the cover of the hive when working with all but our most gentle bees. This is just good policy and helps greatly to keep the bees from becoming angry at us or others in the neighborhood. We approach our bees with an attitude of respect, talk to them as soon as we come near the hives, and visit with them every day so that they get to know their keeper. Even so, some of us do have the strange ability to work more freely and easily with bees than do others, although I am convinced that almost everyone can learn to keep bees.

For many years I have observed a curious fact among our bees, the gentle-tempered bees do not swarm as often as the bad-tempered ones. Also, when the gentle-tempered bees do swarm, they have a tendency to cluster in a high tree where it is difficult to hive them, or they may fly away entirely; whereas the bad-tempered bees often cluster within easy reach and seldom go completely out of our area before clustering. It is easy to see where this will end. If we want to increase our number of hives and hive our own bees as they swarm, we may soon find we have many cross hives and fewer gentle ones.

Some years ago a man of my acquaintance ran this situation to its ultimate conclusion. He bought two gentle hives from me. Then he decided to increase his hives and bought a hive from another man. These last bees were quite cross and would sting without much provocation. Soon they swarmed. He hived them. Later in the season they swarmed again. He hived them. Now he had two gentle hives and three cross ones. Then one of the gentle hives swarmed but they flew off and he could not catch them. Came the next spring. Cross hives Numbers 1, 2, and 3 all swarmed. He caught them all. Now he had eight hives—two gentle and six cross—and they stung him so badly he sold the whole lot to the first man who came along wanting to buy bees. He was through with bees! Later he asked me what he should, or could, have done differently and so have kept his bees—for he still liked honey. Probably he should have immediately re-queened his first cross hive with a more gentle queen but that would have required more experience and know-how than he possessed at that time.

Two seasons ago I happened to stop by to see a young man who has several hives of bees. Two hours earlier he and his wife had tried to work with a certain hive, and both of them had been threatened and stung until they gave up and quit for that day. Then I came along and he pointed out the troublesome hive and told me of their difficulty in handling those particular bees. I saw at a glance that his hive needed two full supers taken off and empty ones put on.

"Would you care to have me work with your bees?" I asked. He eagerly accepted my offer of assistance and we soon had the necessary exchanges made. Neither of us was stung or even threatened.

"Why do you think my wife and I were threatened and stung so badly only such a short time ago, and now we both worked this hive with ease?"

"I don't know," I told him. And truly I didn't. "Did you do everything exactly as you saw me do it?"

"Yes, as far as I can see," he assured me.

"Then I'll tell you what I'll do. This hive will need servicing again in two weeks. I'll come over again at that time and do all the work. We'll set everything out in advance so that you can stay close by and watch me. Maybe next time you'll be able to see what I do that is different from the way you handle your bees."

He accepted my suggestion with a smile of profound relief. Two weeks later when we began to work with his bees there was a gentle cool wind blowing from the south as it had been on the earlier occasion. I smoked the entrance rather heavily because since the hive was facing east, the wind had a strong tendency to blow most of the smoke away before any of it entered the hive to distract the guard bees. Then I pried up the cover a crack on the windward side and smoked the bees several good puffs at that point also. The smoke went right in with the wind.

"I see it now!" my friend exclaimed. "I didn't use the smoker nearly as much as you did and probably didn't get any smoke into the hive entrance at all. And I smoked under the cover from the leeward side instead of the windward side as you did just now." He had found the answer and has had no more difficulty with his bees.

Our "buzzin' cousins" love to be treated to a little honey in the wintertime when there is no honey flow. One recent winter we had a hive that I thought was possibly getting slightly low on stores. I suggested to my father, and he agreed, that I should feed that hive a little honey every day for two weeks. I poured six heaping tablespoonfuls of honey on a plastic coffee can lid where it soon spread all over the bottom of the lid. Then I made a float of little pieces of wood and placed it in the lid to keep the bees from getting "gooied up" in the honey. As I went to the hive, I talked to the bees and removed part of the entrance cleat so the lid could be slipped completely inside. I replaced the cleat to narrow the entrance opening to its usual size so as to give the guard bees an advantage in defending their hive if there should happen to be robber bees in the immediate neighborhood. Bees can smell fresh honey for a considerable distance and, of course, they all want some. In a few minutes I stooped down and saw bees all around the edge of the lid licking up honey as fast as they could.

That evening, with a wire bent at one end, I pulled out the feeder lid. The next morning about ten o'clock I fed them again. As soon as the bees heard my voice at the entrance, some of them came out to see what was going on. When they smelled the honey, they went to work and called others to come, too. The third day a number of bees heard my voice and flew out to welcome me. By the sixth day several dozen bees flew out at the first sound of my approach near their hive, and by the time I could open the entrance wide enough to slip in the lid of honey I found such a mass of bees clogging the entrance that I had to push the lid in very carefully to be sure I did not hurt any of them.

The next day my father came along. As I neared the hive a welcoming committee flew out that was a joy to behold. They buzzed all around me, flew into my face head first so hard that it hurt me slightly, and, in general, made it crystal clear that they were simply delighted to see me. When my father saw me among that swirling swarm of bees without being protected by a veil, he called out, "Ormond, get away from there. They will sting you to death!" It was not easy to convince him that all these beloved bees were just the welcoming committee—putting on a real show for his benefit, I do believe. However, none of them ever did sting me, and much later, long after I had ceased to feed them, they would come buzzing around and check up on me to see if I had a little honey for them. Intelligent? I should say so!

13

Preparing For The Golden Harvest

WHICH SHALL I ATTEMPT to obtain from my bees, comb honey or extracted honey? This is a question that I have been asked many times. Many beginning beekeepers do not want to spend eighty-five to one hundred and twenty-five dollars to buy an extractor, at least not until they have tried their fortune as beekeepers. I can surely see their point. What they usually want to know is if they can get as much honey by putting factory produced four-by-four inch section boxes in their supers as they can by putting in ten extracting frames. In my experience the answer is a definite "No"—at least not here in Santa Cruz. I believe this is because our weather is entirely too cool during our main honey flow. Bees like to work in large groups and by doing so they create heat. It takes high heat in that part of the hive where they are working to exude wax in the first place and form it into drawn comb in the second place. Those little wooden section boxes force the bees to work in small groups. It is my belief that they just cannot generate enough heat to do any amount of work when the outside temperature is seventy degrees or lower. I have heard of no one else in this area who has had any real success with these boxes either. We no longer attempt to encourage our bees to build comb honey in section boxes, but through the years we have sold a large quantity of comb honey. A beginner who would like to try his hand at keeping bees but with the least expenditure of money will do well to use our method of producing comb honey for his own use or for sale. It is true that if we cut out and use or sell the wax comb as well as the honey, we should not expect our bees to produce any large amount of honey per hive because it takes so great an expenditure of time and effort on their part to draw out and replace the comb before they can fill it with honey. Nevertheless, some hives do surprisingly well for us, as comb honey always sells for a good price.

We buy seven-inch paper plates and plastic bags from the grocery store. We take lovely newly drawn and filled combs from the hive, lay each filled medium depth frame down on its side over a wooden box-like frame especially built for this purpose and with a sharp thin knife, cut out one-quarter of the length of honeycomb in the frame and carefully drop it onto a paper plate held beneath. We cut out the next section until all four are cut onto four plates. With a little practice one can do this very nicely and neatly.

We slip each plate with its comb of honey into a plastic bag and close it with a twister. We write the weight in ounces on one of our jar label cards and staple it to the twister that comes with each plastic bag. Now it is ready for the first eager buyer that comes along. There are always plenty of buyers. This gives each customer the pleasure of hefting the various sized combs. Many customers delight to estimate the weight of a comb as they look at it through the transparent plastic bag. They check their guess with the weight on the card or reweigh it on the scale. Some of these people amaze me with their ability to heft a comb and tell how much it weighs. Older men especially can detect a quarter ounce weight variation between two combs, and they want the heavier one, of course. We never try to weigh combs to fractions of an ounce. If a comb weighs sixteen and one-half or sixteen and three-quarters ounces we label it as sixteen ounces. Our customer selects the comb that pleases him best, whether it weighs eight ounces or twenty-four ounces.

Formerly we expended a lot of time and energy trying to make each comb, as it was cut from the frame, weigh exactly twelve ounces. Then one day I found a comb of honey that the bees had drawn out to double the usual thickness. It was so beautiful that I decided to cut it into comb honey and use it for display purposes, as I did not think anyone would buy it. To my surprise, the first buyers that came along each bought one of those huge combs. From that time on we always packaged random sizes and weights, with pleasing results.

Honeybees are such headstrong little creatures that they sometimes completely ignore the full sheets of super foundation we give them and build the combs to suit themselves. It is in these instances that we get curiously drawn combs—extra thick, thin, curved, crooked, or filled with holes. Such combs are difficult to uncap and extract, so we generally do better by setting them aside and cutting them into comb honey. With a little ingenuity we can usually cut these combs in a way that makes them look very attractive.

I really love this comb honey myself. Years ago I was in poor health and most foods did not agree with me. I went to see our family doctor and he asked me what foods I could eat.

"Bread and honey," I replied.

"That's good," he said. "Heavy on the honey, light on the bread. Take these pills and you will live."

I did, and I lived happily. I also began eating comb honey every day all summer. I still do.

Some of our older citizens say that the only proper way to eat comb honey is to crush it with a knife and spread it on bread—eating it honey, wax and all. Most of us cut off a bite-sized piece, chew it awhile and discard the wax. Some who have sinus trouble or catch colds easily and often, say this comb honey is of help in relieving or even eliminating the difficulty. Certain it is that Father and I have not had a cold or the flu in years and I had both, it seemed, all the time during more than four years in the army when I could get no comb honey, or walnuts. I might add that in our family the other great cold and flu remedy when comb honey is not available is to eat walnuts, either English or black. My grandmother used to say, "Crack and eat the meat of four walnuts every day, and you will not be sick alway."

Let us say that it is now April fifteenth, and in two or three weeks there will be honey ready to take off, extract, bottle, and sell. Some further preparation must be made in advance for this happy event. Plenty of money will enable one to get set up quickly with a new extractor and all other necessary equipment. However, if one has an inventive mechanical bent and would enjoy making his own equipment, he can make most of the things he will need. First of all, each beekeeper should have an extractor of his own, even if he has only

Extractor Room

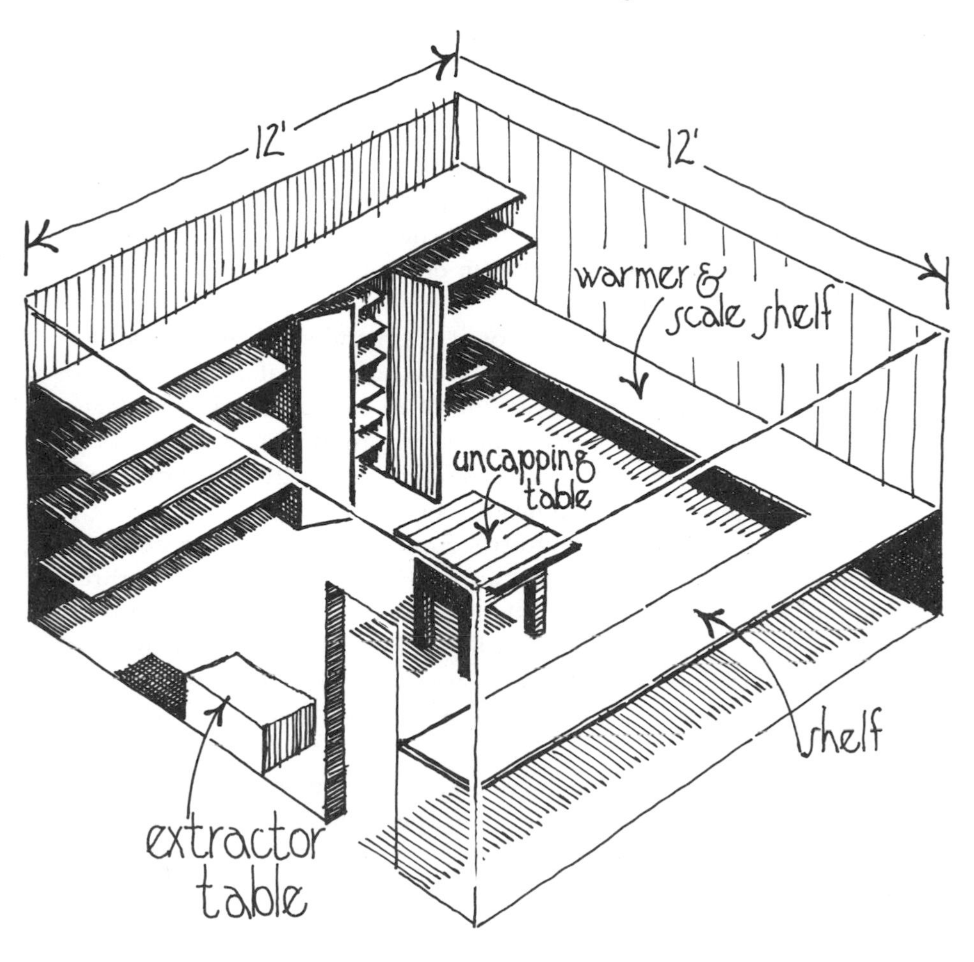

one or two hives, provided of course, that he has proven to his own satisfaction that he can, and desires, to continue to keep bees. It is difficult these days to find someone who will do custom extracting and even so, one is always dependent upon the work schedule of the other person.

To get the maximum amount of honey per hive, we must take off supers, extract, and replace them when our bees are ready. Delays are often costly. A new extractor may have to be purchased but before doing that, try to buy a used one. Often a good used extractor can be bought for twenty-five dollars. To see if it is in good condition check the gears at the top and side. The gears are about the only parts that wear out on an extractor. If it is a little rusty in a place or two, sand off the rust and rub the spot with a small chunk of beeswax. Or, better yet, coat it with a plastic preparation made for this purpose and obtainable from Diamond International Corporation, Apiary Division, Chico, California, 95926; or from the nearest bee supply house in your locality.

Caution: Thoroughly and at once wash and sterilize all used bee equipment purchased. Use plenty of boiling water and soap to cleanse both the inside and outside of an extractor and its parts, as well as every other tool or container acquired from a retired beekeeper. He may be out of business due to deadly American Foul Brood. Safeguard your own healthy bees in every possible way.

It is difficult to build an extractor from available scrap because it is hard to find lightweight steel gears of the correct speed ratio—three to one—three turns of the basket to one of the handle or crank. In one of my antique bee books published in 1888, it states that in those days one could buy these gears especially made for extractors for less than two dollars. Bee supply companies may still stock such parts.

The next thing needed is a screened-off room where the bees cannot follow when one takes off a super of honey. The kitchen can be used as a last resort. At our former location there was a patio with two walls and a wooden lath roof. We built the other two walls with used lumber and window screen, put on six sheets of corrugated aluminum roofing, and had an inexpensive twelve by twelve foot honey room completed in a few hours. To the right of the door we built a two foot wide shelf two feet high the full length of the wall. Filled supers are heavy, so we needed a shelf near the door on which to place them. We also used this shelf to hold the empty supers and frames during the extracting process. To the left of the door we built a low, fourteen inch high, strong table with a three by four foot top. On this we set and anchored down our extractor. Since bees do not always fill the frames with the same weight of honey, the extractor may vibrate badly at times, hence the anchors.

On the far wall to the left, we built shelves to hold clean empty jars and a cabinet with shelves to hold the filled honey jars. On the wall opposite the door we had room for miscellaneous items, among them our honey warmer and jar filler. In the middle of the room stood a table with a two and one-half by three foot top. On this we placed a large rectangular pan. The metal side panel from an old gas range is about the right size. In one end of the pan we built a moveable wooden rectangle of one by six inch boards, nineteen inches long by twelve inches wide outside measure. We have found it easier to rid the room of curious bee friends by covering windows and opening a door, for light attracts bees and causes them to fly out.

When cutting comb honey, we lay the filled frame of honey flat on this wooden frame, which is high enough so that a paper plate can be held under it to catch the cut-out comb, and the big pan beneath catches the honey drip. When we are going to uncap frames of honey for extraction, we place a one by four inch board sixteen inches long across it, lengthwise of the pan. In this board we bore a small hole at about a thirty degree angle somewhere near its center and in this hole drive a wooden peg two inches long and sharpen it on the upper end. To uncap a frame, we stand it with the lower end resting on this peg for a pivot and hold the upper end in the left hand. I am righthanded so I lean the frame somewhat to the right so that as the uncapping knife slices off the caps, they will fall clear of the comb in the frame. We start uncapping at the bottom of the frame and slice up, swing the frame around on its pivot, and slice off the other side. If we want to continue uncapping more frames before beginning extracting, we make a lean-board for the other end of the big pan, lean this first frame against it, and go on uncapping several more frames.

The size of the extractor needed will depend largely upon the number of hives owned. For three hives or fewer, a two-frame, hand-operated extractor will do nicely. But if one has secret aspirations toward eventually acquiring more bees, say as many as twenty hives, then I would advise buying a four-frame extractor. Invariably a successful young beekeeper desires to increase his number of hives as the years pass. This is a praiseworthy goal of which I heartily approve.

Every beekeeper needs an uncapping knife. For a long time we used a very sharp, very thin, butcher knife with an eight-inch long blade. However, by far the most efficient cold uncapping knife I have ever used was one made by my father. He cut it out of an old handsaw blade, shaping the metal so that he could attach a piece of wood at each side of one end to make the handle. He made the blade three inches wide by eight inches long and brought it to a rounded point at the end opposite the handle. He ground and whetted it to a fine edge. This knife had the great advantage of a large flat surface on which to hold the cappings after they were cut from the comb. When the knife surface was covered with cappings, I could scrape them off on the wooden frame. If cut-off cappings fall back upon the area already uncapped they are difficult to remove.

When using the cold-knife method, one will occasionally find a comb that simply will not uncap with a cold knife. In this eventuality we fill a deep kettle with water, set it on an electric plate to heat, put in at least two butcher knives to get hot and when the water is nearing the boiling point, take out a knife, dry it carefully with a clean towel, and uncap a frame. Then put it back in the water to reheat; dry and use the other knife.

Caution: Be sure to dry each knife thoroughly before using, for if even a small amount of water gets into the honey, the honey may spoil.

Of late years we have used an electric uncapping knife. This knife is held at a constant hot temperature that makes it easier to slice through the comb, and it is also partially self-cleaning of wax. A word of encouragement to beginning beekeepers in the use of an electric uncapping knife is in order at this point. It looks so simple to uncap a frame when an experienced operator is handling the knife. It is often quite a different story when a beginner tries his knife for the first time. There is a distinct tendency for both the knife and the operator's temper to overheat badly. Electric uncapping knives have thermostats for heat control; nevertheless they will overheat and melt the wax and scorch the honey unless they are quickly skimmed through the comb on each side of the frame so that the honey exerts a cooling influence upon them. If one uses the knife too slowly, one is forever taking time out to turn off the power and then turn it on again. However, take heart. Practice makes perfect in this as in any other endeavor.

There is one situation in which practice will not help and that is if the thermostat becomes inoperative and leaves the knife on at maximum heat. My knife did this some years ago, and I was about to throw it away when my father suggested that he wire it in series to a 200 watt light bulb above my head that lighted our extractor room. This he did, wiring it so that the current first had to pass through the bulb and then through the uncapping knife. The knife worked perfectly and I am still using it. One time a friend of mine wanted to try his skill at uncapping but as he worked too slowly the knife overheated. I replaced the 200 watt bulb with a 150 watt bulb and that reduced the heat to the right temperature for him.

Next a jar filler is needed. Do not try to fill jars one at a time under the spout of the extractor as the honey runs out. Perhaps it can be done, but I never could do it successfully. Too many times my jars ran over, and there are few things more aggravating than a really 'stickied-up' honey jar. Run the honey from the extractor into a large container such as a four-gallon can. Clean cans of this size can usually be obtained from a small bakery for a nominal sum. Some Chinese restaurants use vegetable oil and the empty cans may often be had free for the asking. These cans must be thoroughly washed before using them for honey. After they are cleansed I always take a thimble-sized piece of beeswax, drop it into the can, hold the can over a gas burner on the stove until the wax melts, and then run the melted wax all around the bottom of the can and also in the seam in the tin running from the bottom to the top. This treatment effectively eliminates any possible leakage in the seams of the can.

If a beekeeper has from one to five hives of bees, the honey can be dipped up from the can and run into the jars with a berry spoon. As the spoon rises from the can, give it a quick twirl to stop the honey drip—then let the honey run off the spoon into the jar. At first a few drops will fall on the outside surface of the jars, but one soon becomes skillful at filling. Mother and I filled many hundreds of pint and quart jars in this way.

Suppose one has been pleased with his bees and, blest with success in buying or improvising the equipment needed, wishes to hive

wood plug
Jar Filler Can & Spout
1½" waste pipe
side view
soldered

more bees to produce more honey. A honey can with a flow-gate valve (shut-off cock) to fill jars is now a necessity, for speed has become essential. One can buy a filler can or make one that will do almost equally well. From a paint factory buy a clean new five-gallon can, cut out the top and hammer down the rough edges. Cut a one and one-half inch diameter hole in one side near the bottom. Find a two inch length of one and one-half inch diameter shiny pipe such as is used as tail pipe under a sink. Solder this pipe over the hole cut in the can so that the pipe sticks out at right angles. To facilitate the soldering, file off the shiny coating on all surfaces that are to be soldered until the brass shows. Take another four inch length of the same diameter pipe and cut a big hole in the side of it a little below the center. Fit the hole in this pipe to the pipe coming out of the can so that this last piece of pipe is in a vertical position. Solder the vertical pipe to the pipe projecting from the can. Now find an eight inch length of straight-grained wood, preferably hardwood, and make it into a long round plug that will slip up and down in the vertical pipe. When the plug is pulled up, honey will flow out the lower end of the vertical pipe. When the plug is pushed down, the honey flow is shut off. In a moment the honey drip will cease for a few seconds and in that time change jars. This is still our present method of filling jars.

Caution: Never use as a container for honey any can that has contained gasoline or any other petroleum product, as it will taint the honey. A thoroughly cleansed vegetable oil can from a restaurant is satisfactory.

Another needed item is a scale that will accurately weigh, in ounces, up to about six pounds to be used in weighing and selling comb honey and beeswax. Before using the scale to sell a product to the public, the law requires us to notify the weights and measures inspector so that he can test it for accuracy. Then he will attach his official sticker-type seal to the scale and also write the beekeeper's name on his list. He will come around once a year to recheck the scale. Long ago I bought such a scale from a grocery store for five dollars. It was old and somewhat beat up, but the weights and measures inspector said it was still very accurate. He gave me clearance to use it, and I have been doing so for many years.

Lastly, combination business and jar label cards are needed. I have been able as yet to buy these cards locally for five dollars and seventy-five cents per thousand. My cards have the following printed on them to meet the requirements set by law in California.

Pure California Honey
U. S. Grade Fancy
Net Weight ______ Lbs. ______ Oz.
Ormond Aebi
710 — 17th Avenue
Santa Cruz, California 95062
Phone 475-2065

Our bee inspector has allowed me to attach one of these cards to each jar of honey with a small rubber band around the jar. This has been of great assistance in selling honey. As the card is small and easily detached from the jar, people put it away somewhere, usually in the kitchen, and even years later, I get letters and telephone calls from these people wanting honey. Such inquiries are always most welcome and gratifying.

Whether or not a beekeeper can have his cards printed "U. S. Grade Fancy" depends upon the results of tests conducted on his honey by the bee inspector. The honey will be tested for purity, clarity, color, density, foreign material, and other factors. "Fancy" is the very top grade. If that rating is obtained—rejoice—and continue to strive to keep all things dust-free and spotlessly clean. California law encourages the aspiring beekeeper to get started and try his luck.

If one has the good fortune to harvest more honey than is required for his own use, he may sell the excess to anyone he wishes without jar labels, and in any type of container, provided the buyer comes to the beekeeper's residence where the honey was produced to make the purchase. It is only when honey is to be sold away from the home place, such as a roadside stand, store, peddling, and the like, that all the aforementioned regulations come into effect. California laws are good and easy to live with and the bee inspector is friendly and helpful with any problems that may arise. I am always glad to see him come around. Our former bee inspector was of definite help to us in our earlier years.

A relatively recent change in the law requires that all hives be registered with the bee inspector by November first of each year. There is no charge in California. Once a year he will come around to inspect all of the hives for disease. There is no charge for this service either. It is always well to check with the local bee inspector to learn the current laws and regulations when beginning beekeeping or when moving to a new location.

By the way, honey is heavy. A pint weighs one and one-half pounds. A quart weighs three pounds. A gallon of honey weighs twelve pounds—net, of course. So fill in the label card as required.

14

Removing & Extracting Liquid Gold

THE LONG-AWAITED day has come. There is honey in the hive in excess of the bees' needs. This honey is ours for the taking! The sun shines brightly, the air is warm, the wind calm. We take off supers of honey in the warmest part of the day for two reasons: First, many bees will be away from the hive gathering pollen and nectar. It is definitely easier to open a hive and remove honey when few bees are present than when they are all at home. Second, we want to conserve as much hive heat as possible during the four minutes the hive is open. To accomplish removing a full super and replacing it with an empty one in four minutes we must have all things ready and at hand beside the hive before we pry up the hive cover. These necessary items include a smoker lighted and going well, extra fuel (pieces of burlap bag), hive tools, bee brush, replacement super, and an extra empty super box with top and bottom covers. The bottom cover keeps the newly removed honeycombs from becoming soiled and the top cover keeps out unwanted investigating or robbing bees. A beginner should slip on a jacket, bee veil, and gloves.

Now with our smoker we give the entrance five good puffs of smoke, being certain the smoke gets into the entrance and is not blown away by a breeze. We never work with our bees on a windy day. We pry up the hive cover slightly so that we can force several puffs of smoke into the hive before any bees can escape; then slowly remove the cover. We give six or eight puffs of smoke across the frames until most of the bees in sight have run down out of our view between the frames. Since bees almost invariably glue all of the frames securely into place before they finish drawing out the combs, we must use a hive tool to quickly pry four or five adjoining frames apart from each other to break them loose from their setting. This enables us to lift up a half inch, one end of the frame next to the center frame and then pry up the other end. When we use drawn frames from the year before we use nine frames to the super instead of the usual ten frames because the bees draw the comb out longer and that makes the frames easier to uncap. We lift the frame out with both hands and look carefully at the honeycomb. If it is completely sealed at both ends and along the bottom rail on both sides, it is ready to take off. We smoke the bees a little from time to time, brush them all off the comb back onto the hive, place this filled comb in the empty super box at hand, and cover it. Now we pry each of the other frames loose and take a quick look at them. If they too are completely sealed, we quickly brush the bees from each frame, fill the empty super with filled combs, and remove the now empty super box from the hive. If there are six or seven supers on a strong hive, we may find that the bees have finished only four or five combs in each of the two top-most supers. In that case we take out these finished combs and combine the raw nectar combs into one super which we leave as top super for the bees to complete.

Our next step is to remove two half-filled supers from the hive (one is our newly combined super), setting them aside for the moment. We pick up the prepared replacement super and place it on the hive, then replace the two half-filled supers on top of the replacement super, add the cover, insulation board, roof, and weights (twenty to forty pounds of weights in winter and spring) and the job is done. A super of honey has been taken off and at the same time the bees have been given room lower down where they need it. With a little practice this can be accomplished very rapidly.

There are other ways to take off honey. One is to use a bee escape board. This is a device made of thin boards built the same length and width as the top of the hive, only it has an elongated hole in the center to receive a metal bee escape. To use it, we check the super to be taken off to see that it is well filled and sealed. We leave all the frames in this full super and raise it up to the top of the hive, placing the bee escape board under it with the round hole side up. By the same time next day all the bees will have left the super by way of the escape and the full super can be lifted off without taking time or trouble to brush the bees from each frame. This sounds simple and indeed it is, but it has limitations. When placing the escape board under a full super, we were obliged to break the seals the bees had made to keep out cold air and also invaders of the hive. Warm air will leak out of the hive until the super is taken off next day, and in our cool climate this is wasteful. Also, in this area, as soon as the bees have left their filled super the tiny brown ants begin to enter through the broken seals, and by the next day we may have a far greater problem ridding the super of ants than we would have had in brushing off the bees.

Sometimes, when we can see at a glance that a super is filled to capacity and the weather is too cool, we take the super off the hive, replace the cover and carry the super a short distance away placing it on a stand made for this purpose. Then we remove the frames, brush and smoke off the bees and take the honey. These bees will soon return to their hive. The advantage of this method is that the hive is opened and closed again in about two minutes.

There are times when we look quickly at the frames in a super and see that the cells are almost, but not quite, all sealed, a few unsealed cells of honey being at each end of the frame. I am often asked if these frames may be removed and extracted and will the honey from such frames keep or spoil. The answer: Take a moment to test each side of each frame. Hold the frame lengthwise over the hive in a flat position. Give it a sudden quick shake or jar. If the honey in the unsealed cells is still too raw to keep, it will shower out onto the frames below. This honey is not lost because the bees will gather it up again in a few minutes. But this frame is not yet ready to extract. Replace it on the hive for another week or so to ripen. Try the frame next to it in the same way. Sometimes one or two frames have raw nectar and all the other frames of honey are ready to remove. In this situation take off those frames of honey that are ready and replace them with empty frames (previously extracted frames) or, if early in the season, frames with newly inserted starter sheets may be put in, leaving the super on the hive.

Through the years I have noticed that beginning beekeepers who did not have too much trouble when they put the supers on often get

Uncapping

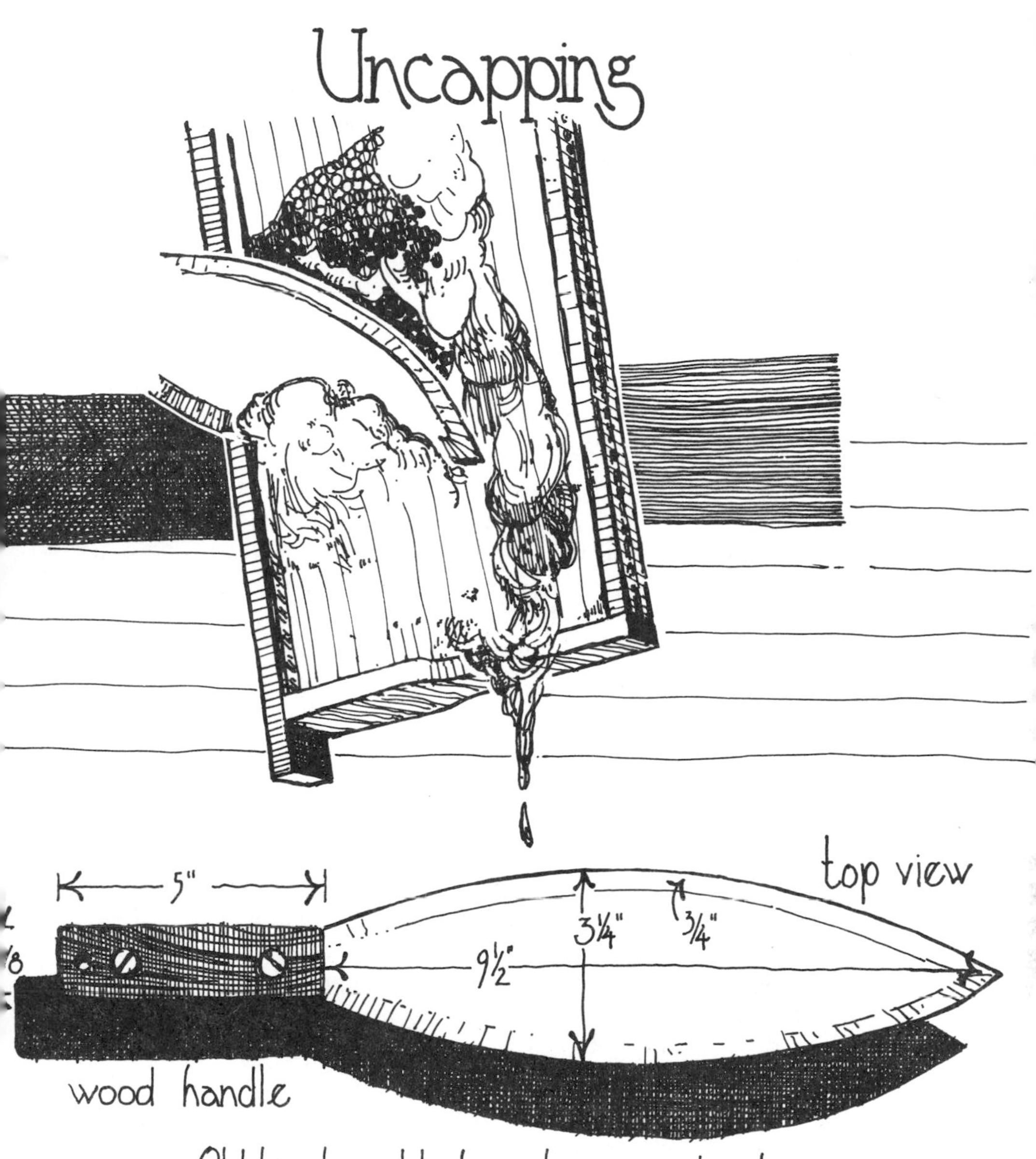

Old handsaw blade · sharpened edges

Home Made Cold Uncapping Knife

badly stung when they take off the first full super of honey. The difficulty began after the hive had three, four, or five supers quite well filled, and they wanted to take one off. I have helped several beginning beekeepers this summer, and I believe we have found the reason. When they put on the empty supers, the hive was still building up rapidly in numbers and very little smoke was required to subdue the guards. When the time came to take a super off, each hive was very strong and the beekeepers should have used much more smoke. This is especially true when removing the cover. We pry up the cover on one side and the back just enough so that three or four good puffs of smoke can be forced into the hive in at least two places. During this time the crack under the cover should still be so small that no bee can escape. If the cover is pried up so rapidly that bees can escape before we have time to smoke them, they will surely be inclined to sting in defense of their home. One cannot blame them for this but appropriate precautions should be taken. There is wide variation among hives as to how much smoke will be needed. Burlap bags, usually obtainable from any feed store, make excellent smoker fuel when cut up into small pieces suitable for filling a smoker. This smoke will not hurt either the beekeeper or his bees. It is often better to use a little too much smoke than not enough.

Caution: If flame begins to puff out of the smoker, add more fuel quickly to keep from burning the bees. We want them to think their home is on fire so they will endeavor to protect it, even as you and I would do, and so pay less attention to what their keeper is doing. But we do not want to singe their wings or hurt our frightened friends.

The rule is that the sooner honey can be extracted after it is removed from the hive, the easier it will be to extract. This is because the bees keep both honey and its wax comb many degrees warmer than our usual outside temperatures. The colder the honeycombs become, the harder it is to get the honey to fly out (by centrifugal force) of the uncapped cells. Father and I take off only as many supers as we can extract by ten o'clock that night. We take off supers in the warmest part of the day and begin extracting immediately. We carry the newly removed supers to our honey house where they are placed on a flat table or counter and covered to conserve heat. We check the extractor to see that the anchor chains or wires are tight. Three light wire chains are hooked over the top of the extractor, one on each side and one at the back, with a turnbuckle fastened to one end of each chain. Each turnbuckle in turn fastens to a hook attached to a low table top. This low table is sturdily built and securely fastened to the wall and floor. A little honey applied to the lower bearing in the extractor below the basket serves as a lubricant. One should never use oil or grease in this bearing. We assemble the extractor and turn the handle a few times to spin the basket to see that it is assembled correctly. Now with a stout string we tie a six by eight inch medium-mesh nylon strainer bag securely to the extractor drain. The bag needs to be large because bits of wax or crystallized honey tend to clog the mesh. Also we slip a sixteen inch square piece of linoleum under the front edge of the extractor to project over the honey drain, strainer, and can top, to keep from accidentally dropping lint or other foreign matter into the honey can. We place a three to seven gallon capacity can under the drain to catch the honey. It is always well to have several extra cans on hand so that none need be filled completely full.

We uncap some frames as previously described. If using a two-frame extractor, we place the frames on opposite sides so they will balance. We always take a quick glance to see if two frames of about the same weight have been uncapped and use them together to minimize vibration. Then we begin to turn the crank at a moderate, easy-to-turn speed. Soon honey will begin to collect on the walls of the extractor. This honey will slowly settle to the bottom and drain out into the can, but it will take the contents of two or three frames before the honey begins to run out. When the first side of the frame is emptied of honey we turn each frame around in the extractor so that the other side now faces out and again turn the crank as fast and as long as before. Then we lift out the frames to see if all the honey has been removed. If not, we lower them back into the extractor and turn a little faster, reverse the frames and do that side again also. Experience will determine how fast and how long to spin the frames to extract all of the honey. However, it is always advisable to turn a little too slowly at first rather than to begin turning too fast and risk breaking the combs out of the frames. Occasionally, when taking a frame from the extractor, we see that the basket was spun too rapidly and that the comb is cracked the full length of the frame. With our fingers we push the empty comb straight again in the frame and, if necessary, tie it in place with several windings of cotton string (nylon or other synthetic strings are too strong for the bees to chew) around the back and bottom rail. We replace the broken comb in the super and return it to the hive. Bees are masters at repairing damaged combs and it is far more profitable for us to give them back a damaged comb rather than just a frame with new starter. When the bees have finished the repair job they will cut and carry out the cotton string.

Caution: Check frequently to see that the honey can is still under the extractor drain spout. Often in passing someone kicks the can just enough so that the honey misses it and runs out onto the floor.

When taking off the first supers of the season, some frames will be found to contain partially crystallized honey. We uncap these frames in the usual way and spin out all of the honey possible, turning as fast as we dare. Even if much crystallized honey still remains in these combs, we put them back on the hive for the bees to rework and they can be extracted the next time around. If the honey is crystallized so hard that none of it will spin out, we lay the uncapped frame flat on its side and from a height of three feet, pour some lukewarm water onto the honeycomb on each side of the frame. The warm water helps the bees rework the hardened honey. This they always do before the honey has any chance to spoil. There is a common belief that liquid honey does not spoil, and this is true; however, it can over the years change consistency between granulation and liquid so many times as to become less palatable.

Until a few years ago we did custom extracting as an accommodation for fellow beekeepers who had only one or two hives of bees. We made no charge for extracting and did all of the work, so it was strictly a nonprofit operation. I did the uncapping and my father ran the extractor. All we asked of the customer was that he bring his own containers for his honey and that he watch closely while the extracting was going on to see that his containers did not overflow. Most people brought cans or jars never larger than quart size. You guessed it. These people became so interested in watching the operation or telling stories while we worked that they completely forgot what they were supposed to do until a container had run over and a great puddle of honey had accumulated on the floor. Then we had to stop the whole operation and clean up the mess. Often a few minutes later the same thing would again occur.

Finally we gave up accommodation extracting and charged a flat rate of three dollars a super. In addition to doing all the work, we also furnished on loan one of our big honey cans into which to run the honey, and sent it home with the customer so he could fill his own containers. This saved much cleaning, time, and temper all around. I am glad to be able to say that these people were always thoughtful enough to return our honey cans in good condition.

People find a peculiar thrill in watching the extracting process. The sweet scent of honey, coupled with free honey covered cappings to munch while the work progresses, plus an ever rising level of gold in the honey can, reminds many of the "pot of gold at the end of the rainbow." Indeed for all beekeepers it represents the culmination of many months of hopeful endeavor and eager expectation. It is a time of general rejoicing for everyone in or near the extractor room.

15

Requeening A Queenless Or Crosstempered Hive

WHEN THE FIRST HONEY of the season is being extracted and we and our bees are working almost day and night to keep up with the wild flower bloom, we may discover that something is wrong with one of our hives. Five years ago this had occurred when we noticed that our Number 10 hive, which was very strong in numbers and carrying to their hive large amounts of both nectar and pollen every day, suddenly became listless and acted hopeless. A few bees came and went, but not many, and they brought but little nectar or pollen. A dozen bees loafed around the entrance without aim or purpose. It was the unmistakable sign of a queenless colony with no eggs or larvae young enough for the bees to rear a new queen. Poor bees! Despair showed in their every act. What could we do?

A cold north wind was blowing so we could not possibly open another hive where there might be a ripe queen cell, take it out, and give it to the queenless colony. So we waited until the weather warmed up but it did not moderate for three weeks. Then we had an unexpectedly warm afternoon. To our surprise we saw a small swarm of bees cluster on a plum tree near our barn. Where it came from we had no idea. It was not from any of our hives. We thought of hiving it but it was a mere two handfuls of bees. This small swarm would never make it anyway—so why bother? Then my father remembered our queenless colony. Yes, indeed! This tiny swarm was sure to have a queen and we had a strong hive that needed a queen most desperately. Joyfully we hived this wee swarm into a super-type hive, removed the cover from our strong hive, placed a window screen cover over it, removed the bottom board from our new hive, and set our precious little friends on top of our strong hive. In half an hour the bees at the entrance and on the landing board began to come to life. Someone had scented the presence of a queen in the top super and was spreading the wonderful good news. The hive bees could not reach their new queen because she was separated from them by the screen. But they knew she was there and how they did rejoice!

We took the screen out at noon the next day and by that time the bees were singing a song of joy and were going happily about their work. They made us several more supers of honey, despite the fact that they had been queenless at the height of the honey flow. I should mention that this hive had no queen excluder to keep the new queen from going below into the brood chamber. Had there been a queen excluder on the hive we would have had to handle the combining differently to make sure the queen had access to the brood chamber.

Someone might ask again why we place a screen between the old hive members and the newcomers for twenty-four hours. It is because if placed together immediately, the workers of different swarms often fight desperately and in the melee the queen may be killed. Always play it safe.

Why do hives sometimes suddenly become queenless in this way? I do not always know. One warm winter day I opened a hive to see how the bees were faring and to my delight I saw the queen right on top of a frame for me to look at and admire. She stayed there for half a minute or so. Then, to my dismay, she took wing and flew away over the top of the barn. I waited anxiously for her to return but she did not come back. Sadly I closed the hive and told the bad news to my father. There was no possible way to requeen at that season of the year so slowly but surely the hive dwindled in numbers and finally died out. A meddling beekeeper is said to be the bees' worst enemy, and this may be true. It surely was for that unfortunate hive.

If everything has been done correctly as far as is humanly possible and we still have bees that threaten to sting us whenever we come near, such a hive should be requeened. If one of our gentle hives appears ready to swarm soon, we wait for it to swarm and requeen the cross hive by combining them in the manner described earlier and as follows.

Caution: Always try to do the requeening as early as possible in the spring before adding too many supers on the hive. The fewer the supers, the easier it is to do.

Suppose we want to requeen a cross hive but cannot obtain a gentle tempered swarm until about May twentieth, at which time the cross hive consists of a hive body, a queen excluder, and six supers above the excluder. We hive a new swarm, take the top cover off the established hive and put on the window screen cover with the new swarm in its bottomless hive on top, leaving it there for two days (forty-eight hours). The usual time early in the spring is about twenty-four hours, but now the bees must remain penned up longer because the new swarm is so far removed from the brood chamber. Two days later we remove the top hiving super, using the screen as a bottom so the bees cannot as yet escape and using considerable smoke below the separating screen to subdue the cross bees in the established hive, we set the hiving super aside, then remove all six supers (two at a time if we can lift them) and set them aside.

Now using smoke as needed—a good deal if the bees are really cross—we pry up the queen excluder and remove it, take the swarm caught two days previously and set this hive over the open brood chamber. Slowly, and with some smoke, we remove the screen separating the two swarms. It helps a great deal to have the assistance of a partner. We remove the top cover from the newly added hive and smoke and brush the bees down into the brood chamber. This must be done carefully, for we do not want to injure the new queen. As soon as all the bees from the new swarm are smoked and brushed down into the brood chamber, we remove the now empty temporary hive. The queen excluder is put back on and as quickly as possible we replace all six of the supers that were taken off. We replace the cover, insulation board, roof and weights—and the job is done. This sounds a bit difficult and as a matter of fact it is.

If both old hive and new swarm are large, the combining must be handled in a somewhat different way. First of all, we hive the new swarm in a regular full depth hive body. We place this hive on top of the established hive with the screen between as already described. After forty-eight hours we remove the screen-bottomed new hive, set it aside and remove all six supers and the queen excluder. Now, using some smoke, we remove the screen bottom from the new hive and place this hive directly on top of the open brood chamber. We do not smoke or brush the bees down but leave both new hive body and brood chamber as one unit because a large number of bees requires this much room. We replace the queen excluder on the now double-deck hive and again replace all the supers on top with all possible speed. We always plan to have some full depth frames drawn by the bees the year before, so that when combining in this way the new queen can immediately lay eggs in both parts of the hive or the worker bees can use the drawn combs for storage purposes. If much smoke had to be used to subdue the cross bees in the brood chamber, a great many bees may have gone out the entrance and hang in a cluster from it. This is all right for they will go in again after an hour or so.

Caution: Always remember in all requeening not only to remove the screen cover from below the new swarm but also to temporarily remove the queen excluder and work the new bees down below it. Otherwise the new queen will be up in the honey supers, and the new swarm drones will be penned up there to die, as they cannot get down through the queen excluder to freedom. A pungent rotting smell coming from the hive entrance is sometimes a startling reminder that we have forgotten to liberate the drones.

If we have only one hive and it is cross, we must get a swarm from somewhere else before we can requeen. We have had no success in letting a cross hive swarm even two or three times and then taking each new swarm and recombining it with the parent cross hive hoping to get a more gentle strain, the theory being that a young queen from a cross hive may have mated with gentle-strain drones and so produce more amiable progeny. For us this procedure just does not seem to work. So we let it be known among our friends, the fire department people, or even a roofing concern, that we want a swarm of bees. As soon as we are notified of a swarm, we go hive it and combine it with our cross hive.

Someone may well ask, "Is most any such swarm likely to be more gentle than the one I already own?" The answer is yes, at least in our experience. When the gentle bees swarm, they often fly completely away out of the general area of their keeper and so are made available to others. And since, in our experience, the queen of the new swarm is always the victor in the battle of queens that follows combining, the result is that within thirty days we have a more gentle hive. In a period of ten years we had all of our hives so gentle that I could work any of them without veil or gloves and never get stung unless I accidentally squeezed a bee while rapidly taking out frames of honey. Even in doing this, one gets to know the touch and feel of bees between fingers, and learns to let go or lift a finger before the bee can sting. The result is that of late years I have rarely been stung.

When working with my bees, I do often wear a veil and would recommend that others do, too, for the following reasons: If a bee gets tangled up in our hair, any of them will be inclined to sting because they are frightened. We make the matter worse by taking a swing at the bee or by trying to brush it out, whereas it might extricate itself in a moment or two if left to itself. Again, if one wears glasses, use a veil. If a bee feels trapped between our glasses and our eye, she will likely sting where it hurts the most!

But a honeybee does not want to sting us from a cruel desire to inflict pain upon us; for each worker bee knows that when she stings it means her own life must be forfeited. Honeybees sting from desperation, not vindictive pleasure. So let us do everything possible to preserve the lives of our friends the honeybees.

16

Honey In Abundance To Pack & Sell

As of May first, we have four hives and each of them has from four to six supers above the queen excluder. The top super on each hive is filled and sealed. The lower supers are well along toward being filled, with bees hard at work at every level. We may remove and extract the honey from the topmost supers as there will still be a goodly golden store on each hive to encourage our bees to continue gathering nectar. It is obvious that the season as a whole will produce more honey than we will need for our own household use as well as emergency winter rations for our bees, so we are free to sell some honey at this time to our friends and neighbors. The long awaited happy day has come.

We remove the filled super from each hive and put an empty one in its place. If we have no empty supers available at the moment we replace the hive covers, extract the supers taken off, and put them on again next day beneath the two topmost supers of each hive. Uncapping honeycombs, the first step in extracting, is always a sweet job, for as the work progresses everyone present may fill his mouth with freshly cut cappings. This honey, chewed slowly from the wax cappings, is indeed delicious!

The pleasant task of uncapping and extracting is now finished. We have eight gallons of honey in two five gallon cans—four gallons in each. We set these two cans on the table or counter and cover each with a clean white cloth. There they remain until the next afternoon or a few days longer if need be. With a big spoon we skim off the whitish foam made by air bubbles rising from the stream of honey as it poured out of the extractor drain through the strainer bag and into the can. Now we stir each can thoroughly with a clean one by one by twenty-four inch long stick previously prepared for this purpose and pour half the contents of each can into the filler can, and again stir the honey in the filler can with the stick. We pull the plug and fill jars until all the honey remaining in both cans may be poured into the filler can. After stirring again, we fill jars until all the honey is packed and the filler can is empty. We never fill the honey jars quite full but always leave one-half inch head space. If a jar is overfilled, the lid will be forced off when the honey granulates as granulating honey expands with astonishing power like ice and our customer may have a gooey cleanup problem later on if he does not use some honey out of the jar before it granulates.

Caution: Thorough stirring of honey before filling of jars is essential to success because as honey comes from the extractor and then stands until the next day to be skimmed, it has a tendency to settle. This causes thin honey to be on top and extra heavy honey to be at the bottom of each can. A thorough stirring mixes the honey to its original consistency.

In recent years our bees produced so much honey that we could no longer handle the extracting operation in our honey sales room. To provide more space we rebuilt the loft of our barn and made an extractor and honey storage room out of it. This room, being upstairs, was many degrees warmer than our honey house had been. We could take off additional supers and extract them more quickly than before because the honeycombs kept warm hours longer. Here we made another discovery. If we filled our jars with honey and left the lids off for two weeks, the honey in the jars had a more delicate flavor than before. Agreeably pleased, we built a dustfree cupboard and stored most of our honey in it for two weeks or longer to "ripen," as we called it, in the warm air. Sometimes these jars would begin to granulate, and if so, we left them untouched until the granulation process was fully complete. People loved this specially treated honey—either liquid or granulated.

On occasion we had more call for liquid than granulated honey, making it necessary to return some jars to liquid form. This was easily done with a honey warmer. We built one that would liquefy twenty-five quarts at one time. Such a warmer is a big wooden box built twenty-two inches square, inside measure, built of one by twelve inch lumber. It has a plywood bottom covered with two inches of glass wool insulation on the inside. Some type of insulation board should be placed around the inside walls of the box, as well as the cover. The more heat-proof the warmer is constructed, the less time and energy it will take to return the jars of honey to liquid form. We arrange an electrical system on the glass wool insulation covering the bottom of the box so that we have five twenty-five watt light bulbs glowing at the same time arranged equal distance from each other and from the walls of the box. A false bottom of very heavy screen or small rods is placed above the light bulbs. On this bottom we place twenty-five jars of honey. When the electrical system is plugged in, all five light bulbs should light up and throw light and heat up and around the jars. After six hours we take off the insulated cover and look in. Some jars will be beginning to liquefy. If some are well along and others less so, we rearrange the jars in the box, close it again and look in every two hours. In ten hours' time all the jars should be liquid. Five watts per jar or five twenty-five watt bulbs for twenty-five jars is heat enough. Honey retains its exquisite flavor by very slow warming. We remove the jars from the warmer and immediately close them with caps and rings. The cooling honey contracts slightly causing a partial vacuum in the headspace between the honey and the lid.

Now comes the question of which jars to use. New jars may be obtained in many sizes and shapes. The thinner the glass and the flatter the jar, the lighter in color the honey appears to be. What would seem to be dark colored honey in some round three-pound quart jars would appear as much lighter honey in a flat one-pound jar. But most folks are well aware of this fact and do not hesitate to buy honey in round pint or quart jars. In our county we are required to pack all honey in glass containers, but I understand this is strictly a local regulation. However, it is a good practical policy for all small beekeepers to pack in glass containers. People love to hold a jar of honey up to the light and look through it. This makes the honey almost sell itself.

New jars are quite expensive. This led me years ago to check into the possibilities of packing honey in used pint and quart fruit jars.

The honey inspector said I might do so, provided I could get the jars so sparkling clean he could not tell them from new ones. This I succeeded in doing by using dishwashing detergent and concentrated effort.

After the year's honey is all packed and sold I spend a good deal of time looking for fruit jars. There is a certain thrill in this search. The old saying, "Gold is where you find it," is true regarding used fruit jars. I never know where I will find someone either in town or in the country who has a large quantity of jars to sell. Sometimes I find old and rare fruit jars among them. For instance, last winter I discovered a woman wanting to sell twelve dozen jars with all kinds of fruits and vegetables still sealed in the jars. I bought them all. The berries, cherries, jams and jellies were perfectly good and we ate them. Canned vegetables we always throw out. When the jars were all emptied I found I had eight ancient blue quarts, some blue pints, and several "California Mission Bell" pints and quarts, plus some other oldies and oddies. I now have quite a collection of such jars. If a beekeeper is at all antique-minded, he may have some fun this way, too, as well as a bit of profit.

Caution: Never buy jars that have a white circle around the inside, showing that some liquid has evaporated out of them. Such white streaks will not wash off. You may have to buy them in order to get the whole lot but do not think you can use them to pack honey.

It is difficult to get the rusty lids and rings off these old jars without chipping the glass around the top. My most successful method is to lay a screwdriver with a thin blade flat across the top of the jar. I push the blade under the ring on the opposite side and twist the handle a little to right and left to force up the edge of the ring. Sometimes I have to work the blade completely around the ring before I can take it in my hands and unscrew it, but at least I never chip a jar this way.

Selling the year's production of honey is always an interesting and challenging endeavor. Our goal is to encourage the public to come to our ranch for honey, but it always takes time and effort to work up a clientele in any business where one has a product to sell to the public. This is equally true in the sale of honey. The most interesting and direct way that I have found is to load my car with about twenty quarts of honey and twenty honeycombs and go out peddling for an hour or two. I can just hear someone say, "Oh, I wouldn't go out peddling!" Why not? It is exciting and fun! People love to talk to a

Opening a Stuck Jar

peddler and see what he has for sale—especially if he has honey. Carrying a six-pack tray of honeycombs balanced on my left arm and a carrier of four quarts in my right hand, I head for the nearest place where people congregate—usually a shopping center. I go into one small place of business after the other to talk and sell honey. It is so unusual to see a honey peddler these days that it is not long before I am surrounded with happy interested people.

The honey sells itself, for a jar of honey held up to the light, or a comb of honey held in the sunshine, appears attractive and tempting to the point where people find it irresistible. The weary shopkeepers welcome my coming and invite me to come again for my visit provides a pleasant interlude in the day's routine and leaves everyone in a pleasant frame of mind. I never stay more than a few minutes. Beauty parlors, barbershops, service stations, garages, cleaning establishments, flower shops, and machine shops, are among my best customers. However, anywhere and everywhere that one can find people one can sell honey. This is true even in admittedly difficult places such as mobile home parks. These parks are almost invariably posted with "No Peddlers Allowed" signs. Such signs excite a mischievous streak in me and I am tempted to see if they really mean what they say. Sometimes they do. On one such occasion I drove into a park and stopped in a prominent location. Leisurely I loaded up my carriers and then walked slowly up the nearest street. Suddenly a human bulldog jumped out at me from between two mobile homes and barked, "Get out!"

"But I have delicious honey—."

"Get out! I don't care what you have!"

I smiled sweetly at him and stood still. Then he said, "If you go another foot up this street I'll throw you out on your d----- ear."

I replied, "Sir, I don't have that kind of an ear. All I have is really delicious honey. Wouldn't you like to buy some too? Everyone else does." And I held up a honeycomb to the light. For a tense minute his hostile gaze shifted from the honeycomb to my face and back again to the honeycomb. Presently I saw him begin to relax. He had taken the bait and I knew I had won him for another customer.

"Come on up to the house," he said, "I *do* want some honey. I haven't tasted good comb honey in many a year." When I left, he and his wife invited me to stop whenever I passed by. This I did over a period of years.

Many towns and cities have ordinances prohibiting peddling without a license. The usual fee is fifty dollars a year which is, of course, utterly prohibitive for a short season honey peddler. I always try to respect these laws. However, there is no law stopping a person from delivering honey into these areas. So I make a few telephone calls to friends and acquaintances in such a restricted area and get an invitation from them to come to their homes to show my wares. Then I park a few blocks away, load up my carriers and rather deliberately make my way toward the home of my customer. Along the way I meet as many other people as I can, and when they express an interest in what I am carrying, I always stop and talk to them. Soon people ask if I have any honey I could sell them and I answer that I do. It would be immoral for me to say I do not when in reality I have plenty, and it would be most impolite, even rude to ignore these inquiries, don't you think? So we do a little business to the satisfaction of all concerned. I am not a peddler at this time, just a delivery man.

One morning I was called to deliver a quantity of honey to a business establishment directly across the street from the city hall of a small town that I knew had an ordinance against peddling. I really intended to make just that one delivery as quickly as possible and drive on. But I had to park across the street from my customer and while crossing the street, an old acquaintance of mine, the city building inspector, called out to me, "Hey, what are you doing here and what do you have to sell?"

"Honey," I replied, "just making a delivery."

"When you are through, bring me a quart," he said.

This was something I had not expected and it was definitely risky. But a dollar is a dollar and he had asked me. I had not accosted him to sell honey, so I agreed. I met him in front of the police station. He was delighted with his honey, and when he had paid for it he said, "Wait here just a minute and I will step in and see if the police chief and his staff want honey, too."

I sincerely hoped this time that no one would want honey—but they did—for in a minute he was back with an invitation to come in and bring the honey along. I did. We had a nice talk about bees while almost everyone present bought honey. As I turned to leave, the police chief said to his secretary, "By the way, don't we have an ordinance in this town requiring peddlers to have a license?"

"Yes, we do," she replied, "but you invited him to come in and we have all bought honey from him before anyone thought to ask whether he had a license—or even tell him he needed one."

"Hmmmm—," said the police chief.

That was my cue to thank them all most heartily for their patronage and then make a hasty departure from those parts while the going was good. They let me get away in peace, bless their hearts!

17

Midsummer Care & Robbing

In an area of exclusive wild flower honey production such as Santa Cruz, we never add a new super—that is, a super with new frames and sheets of wax starter—and expect the bees to draw out the combs after June first. It seems that even a strong hive cannot make wax to any extent after that date. This applies to both new swarms hived earlier in the season and established hives carried over from prior years. Late swarms caught after June first are able to draw at least some honeycomb for their own storage needs. Therefore about June first, we check all our hives and take off any undrawn frames, for if they are left on much longer, the bees will cut out sections of the sheets of starter foundation and use the wax for sealing cells and other purposes.

An elderly friend of mine placed a new super on each of his hives about July fifteenth, thinking that surely at least some of his bees would draw and fill part of the combs. However, when we opened his hives on September twenty-seventh, we found that none of the bees had drawn any comb and that to a considerable extent the starter comb had been damaged by the bees carrying away part of the wax. We took off all those empty supers, and he stored them away for early use next spring. To my surprise his bees were still gathering small amounts of nectar from a source unknown to either of us, but were storing it in those frames that had previously drawn

Robber Cleats

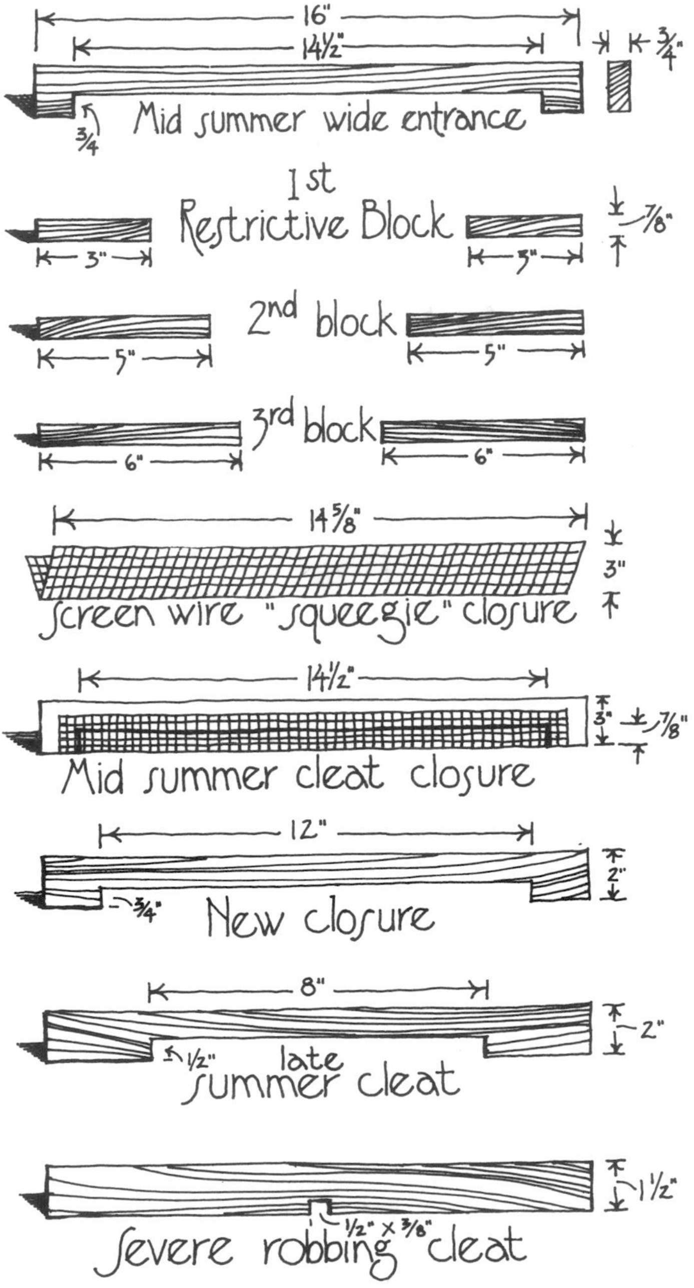

16"
14½"
¾"
¾
Mid summer wide entrance

1st Restrictive Block
3"
3"
⅞"

2nd block
5"
5"

3rd block
6"
6"

14⅝"
3"
Screen wire "squeegie" closure

14½"
3"
⅞"
Mid summer cleat closure

12"
2"
¾"
New closure

8"
2"
½"
late summer cleat

1½"
½" x ⅜"
Severe robbing cleat

combs. The bees were not finding enough nectar to make any excess honey for him but they were adding to their own honey supply to meet their winter's needs.

To find out if the honey flow has in spring begun, or in fall ended, we place a tablespoonful of honey in a pie pan or other shallow container out in the sun where our bees have access to it when they see or scent it. If there is a substantial honey flow, only one or two bees, if any, will come to our pan. But if nectar is unavailable bees will find our offering of honey usually within fifteen minutes and will call others to come to the feast.

Our local honey flow, for all practical purposes, ceases after July first. From then on the bees use as much honey as they gather, frequently even more. If the honey flow stops suddenly, the bees may become unhappy and begin robbing from each other, so it is necessary to observe their daily actions closely. If we see what appears to be one bee fighting with another bee and they pull each other and tumble about, a robber bee may have gone into the hive and a guard is trying to force it out. If, during the next few minutes, we see several other bees engaged in like combat, it is an indication that robbing has begun in earnest. To aid the guard bees in their defense of the hive we restrict the entrance and make it smaller with what are called entrance cleats. These are little boards that fit across the length of the entrance and would close it off entirely were it not for an opening cut on the underside of the cleat. The purpose of the cleat is to make the entrance smaller but still allow the bees to come and go. This smaller opening is now more easily defended by the guard bees. In cases of severe robbing, where a dozen bees are in combat all over the extended landing board, we must restrict the entrance so that only one, or possibly two bees, can enter or leave at one time. In a day or two the robbers will give up and we can begin to enlarge the entrance. But we keep close watch on that hive for some time to be sure all robbing has definitely ceased for good. A strong robber force can overcome the guards of a medium sized hive and steal all the honey and kill many of the hive bees in as short a time as two hours. My father and I use small one by one inch wooden blocks of various lengths as a handy aid to constrict or enlarge the entrance cleat opening in times of robbing.

Early prevention of robbing is the best policy. This is particularly true for swarms hived earlier in the season. After July first in our area when the number of bees returning loaded to the hive entrance drops below ninety per minute we place our first restrictive cleat—the one with the widest opening. When the number drops below sixty bees per minute we use our second more restrictive cleat. At forty bees per minute we close the entrance with a cleat and wooden block closures to an opening one-half inch high by six inches long with all but the center one and one-half inches of that opening being covered with robber bee prevention wire screen or one-eighth inch wire mesh hardware cloth. Of course if we have a season of unusually hot weather and see our bees crowding the entrance or covering the landing board because of the heat in the hive, we temporarily remove some of the restrictive blocks and cleats. On a cold or windy day we use several of these cleats, each one with a more restrictive opening than the last one, to extend the hive doorway into what we call an "igloo" entrance. The actual hive entrance being farther removed from the brood chamber conserves the heat of the hive and also retards robbing.

All of us, old beekeepers as well as beginning beekeepers, need to exercise care to avoid thoughtless acts that will result in honey robbing. Honey is a sticky substance and difficult to thoroughly wash from all equipment after we have extracted supers. It takes less effort to set everything outside near a beehive and let the bees clean up. In a few hours they will do an excellent job and at the same time glean considerable honey that would otherwise be wasted. So why not let the bees do it? Early in the season the bees will seldom cooperate because when there is abundant wildflower bloom the bees prefer to gather nectar. Later in the season when nectar is not so plentiful, the bees will do an excellent job of cleaning up. But late in the season, during August and September, we should never set out anything for the bees to clean, for they will finish their job and then most likely go on to robbing.

An experienced beekeeper recently brought this sad fact again to my attention. He had done a little extracting and decided to let his bees wash up. With practically no nectar available, his eager little helpers accomplished a hurry-up job. He was delighted. Next day he noticed an unusual amount of activity around one of his hives but did not stay nearby long enough to actually note what was going on. In the evening he passed this hive again and saw numerous dead bees on the entrance board and on the ground but no live bees anywhere in sight. A week passed and during those days he saw but very few bees entering or leaving this hive. Then I stopped by for a visit and he asked me to examine that particular hive with him. We opened it and he was startled at what he saw. The hive had been completely emptied of honey. In the process the defending bees had put up a terrific fight until at last they had been overpowered and killed. Their dead bodies were in heaps on the bottom board. The invaders had cut and torn the newly drawn honeycombs all to pieces in their greed and stolen every vestige of honey. The whole interior of the hive was ruined beyond repair. This was a new hive and he had only a few months earlier put in all new sheets of wired foundation. He had caught an early swarm to occupy the hive and they had done a fine job of drawing and filling the combs. Now all was in ruin. He had not only lost a hive of bees but also five dollars worth of starter foundation. I told him not to feel too badly for we have all had the same sad experience.

For several years we had some unusually shiny black-tailed bees try to rob our hives. We could never find where they were from, but they would arrive every year sometime in July or August and continue to rob and pester some of our hives until late October. They never came in great enough numbers to defeat a strong hive, but they kept our new or weaker hives in a constant state of agitation day after day.

My father spent much time and thought in trying to outwit these black robbers. One day he hit upon an idea that really worked and at the same time gave him hours of amusement. He made a restricting entrance cleat with an opening one-quarter inch high by eight inches long. Then he tacked a strip of window screen completely across this opening and cut out a little one inch wide section of screen near the center of the cleat so the hive bees could pass in and out. This small opening was ample for the hive bees because the time was early October and there were far fewer bees in the hive than there had been in April and May. The hive bees found this center opening in the screen without difficulty but the robber bees, who knew they were guilty as sin, always tried to sneak in at one end of the entrance opening. A robber bee came flying from wherever it called home, alighted on the side of our hive, ran around the corner, and bumped its head into the screen. Then it flew around the back of the hive to the other side of the entrance and again bumped its head on that end of the screen. Each robber bee that came to the hive took its turn in repeating this act over and over. It was amusing to watch their frustrated antics. The hive bees came and went as usual but the robbers never figured out how to enter the hive and most of them gave up their robbing in a few days or a week. Some of them were very persistent. They were confirmed robbers and made no attempt to find nectar or pollen.

Last fall our No. 5 hive was being bothered continually so I finally swatted and killed three robber bees and that ended the trouble. Once in awhile we have to resort to capital punishment as the bees do. But we have never had trouble with robbing as long as there was a good honey flow.

18

Feeding New Or Needy Hives

ON SEPTEMBER FIRST we begin taking off all remaining excess honey, leaving each hive twelve quarts, or thirty-six pounds for the bees' winter food. In our Santa Cruz area bees fly almost every day all winter long. As a result they eat more honey than if they went into semi-hibernation as in colder climates.

As we look through the supers we see some frames that are still unsealed. We check them for raw nectar by holding the frame in a flat position over the hive and giving it a hard shake. If nothing comes out of the uncapped cells, the honey is ready to extract and will keep. It was unsealed because the bees had no more honey to store in the cells, so they left them uncapped. We rarely see half-filled cells that are capped and this is a blessing for us as beekeepers because sealed half-filled cells are difficult to uncap. When we do find such cells, the uncapping process is both time-consuming and very wasteful of drawn comb because we are compelled to slice deeply through the partially filled sealed combs in order to extract a small amount of honey. We find some frames with raw nectar that showers out as we shake. We separate the uncapped frames that have ripened honey from those that have even a small amount of raw nectar in them. Almost all of the frames at this time of the year have large areas of capped cells. We extract the unsealed combs with honey in the usual way, then extract the raw nectar frames into a different container and feed this raw honey back to the bees. There are almost always a few hives that could use more stores for the winter so we feed it back to them.

Occasionally, however, there are no needy hives. In this situation we want to keep the raw nectar separated from the capped sealed honey. We cannot uncap that part of the honeycomb that is sealed because if we do, all of this thoroughly seasoned honey, as well as the raw nectar, will spin out and be mixed together as it runs out of the extractor. Such honey will have to be fed back to the bees or it will spoil. Therefore, we take the raw nectar frames and before I do any uncapping, my father runs them all through the extractor just as they came from the hive, turning at a rather slow speed. All of the uncapped raw nectar cells empty quickly. Then he returns these semiextracted frames to me and I uncap them in the usual way. After he has run all the frames through the extractor the first time, he waits twenty minutes for the raw nectar to drain from the extractor and then changes honey cans. Often only a pint or two of really raw nectar which must be fed back to the bees will be found in the can instead of the usual two gallons of mixed honey and raw nectar. Then he runs my newly uncapped frames through the extractor. This honey will keep and may be packed.

There are various kinds of bee feeders on the market—both outside feeders and inside feeders. The one I like best is one I made myself out of an old aluminum electric frying pan. I cut off the handle and all excess parts except the ring on the bottom and smoothed it down nicely with a rasp. Then I made a float of small wooden slats with three-eighths inch spaces between each slat. The float loosely fits the inside of the pan when the pan is empty. On the bottom of the float is a small nail in each corner extending one-quarter inch. These nails support the float so that the bees can easily crawl beneath the slats to clean the pan as the honey lowers. The purpose of the float is to keep the bees from drowning in the honey when the pan is full. At this time of year we need every last bee and cannot afford to lose even one.

When using this feeder, we take out the float, pour two quarts of raw honey into the pan, then replace the float. We take the cover off the hive we want to feed, place a super box without frames on the hive, and lower the filled pan onto the frames below. The circle of aluminum containing the electrical wiring keeps the pan from

Entrance
Feeder
float
3/4"
3½"
5¼"
Double pointed tacks in each corner
½"
deep
6½"

crushing any bees which may be below the pan when it is set down. We replace the hive cover. Ordinarily, by the next afternoon the bees have all the honey taken out and the pan is nice and clean ready to be removed. If we are feeding a new swarm and the bees have but little drawn comb and we have no full depth combs to give them, then it is best to crush some of the newly drawn and filled raw nectar combs and feed it all, both nectar and wax, to the bees. In two days when we look in again, they will usually have removed not only all of the raw honey but also much of the wax as well. They used it to build additional storage space for their needs.

"When you feed a needy hive in late fall or winter due to lack of sufficient stores, or to stimulate the queen bee to begin laying eggs, do you give your bees crystallized honey or do you heat it to return it to a liquid form?" This important question has a simple answer. We always feed crystallized honey. We usually store our feedback honey in widemouthed pint or quart jars until needed. Then we remove the jar lid and with a sharp-edged tablespoon scrape the hard honey a little at a time until it becomes pliable. After a considerable mass is scraped free we empty it into our feeder and scrape more until our entrance feeder pan or larger inside pan has a sufficient quantity. In one or two days, depending upon the amount given, the bees will lick the feeder clean and store the honey in the combs.

Beginning beekeepers who have several needy hives often ask me if they can substitute sugar water as feed for their bees instead of buying pure raw honey for them. The answer, as my father and I have found through the years, is both yes and no. Sometimes we have been quite successful feeding sugar water—and then again we have failed and lost a hive. Some bees will very readily take sugar water while others will not. This seems to be due to the strength of the hive and the season of the year, as well as other factors as yet unknown to us.

One year we caught two very late swarms. They were both large. We hived one into a hive with all drawn combs and fed them sugar water using my big pan feeder. They thrived. We put the other swarm into a hive having only two fully drawn combs, as that was all we had available. We pan fed this latter hive with two quarts of sugar water the same as we had done the first hive. Two days later they had taken only part of the sugar water so we let the feeder remain in the hive. Two weeks later, when we looked in again, the sugar water was still there. I tasted it. It had begun to ferment due to the heat in the hive. We discarded the sugar water and filled the pan with honey. The bees used it as they had need and stored the balance as they had available space. Now they, too, began to thrive. Had we continued to use sugar water, I am quite sure we would have lost them before spring.

On occasion we do use sugar water during the month of December to stimulate a weak hive so that the queen will begin laying eggs for the spring buildup. To make this preparation, we use one pint of 120 degree water and one pint of dry sugar. We pour the sugar into the warm water and stir it until it is all dissolved and the water looks clear. Then every morning, after the sun has warmed the front of the hive, we pour one-third cup of the warmed mixture into a lid feeder and push it into the entrance of the hive. After dusk we remove it using a wire bent one-quarter inch at right angles on one end. If some sugar water still remains in the lid, we give the bees a little less the next day. If we have honey, we give each needy hive one-quarter cup each day. We always prefer to save and feed honey rather than sugar water for winter and spring needs.

On or about November fifteenth we give each hive a dose of the drug Terramycin. Its primary use is as a preventive measure for both European and American Foul Brood. But it also seems to have a needed and beneficial tonic effect upon healthy bees. In years past, Terramycin was packaged in four ounce (and larger) glass bottles packed into small cardboard boxes. The outside of each box had the picture of a calf, a lamb, and a pig. Inside the box, wrapped around the bottle, was an instruction sheet giving the dosages for the various animals including honeybees. Today I find Terramycin packaged in small plastic bags with the usual dosage instructions for all the farm animals except bees. The drug formula is the same as it always was so plastic bag Terramycin is satisfactory for honeybees. The dose for one hive is *one teaspoonful* of Terramycin to one ounce of powdered sugar. Mix the dry ingredients thoroughly and then scatter the light-orange colored mixture over the ends of the brood chamber frames below the queen excluder. A simple and more effective way to feed the Terramycin, in my experience, is to pour the dry mixture into a jar lid and slip it into the hive entrance at midmorning on a sunny day and remove it the next day. If some of the mixture has hardened due to moisture in the hive, crush it to a powder and give it back to the bees for one more day. One dose a year is all I ever give my hives.

If one has, or suspects he has Americal Foul Brood, then the instructions call for two more doses to be given each hive at five day intervals. But in this eventuality by all means call the bee inspector without loss of time. He will inspect your hives without charge and advise you what to do.

If one plans to give bees Terramycin the law requires him to do so at least four weeks before the beginning of the honey flow. That is why we treat our bees the middle of November. Most feed stores that sell hay, grain, and other farm and animal supplies also have Terramycin. The drug has a time limitation factor so always ask for or look for the expiration date on the packaged Terramycin.

In 1972 we were presented with a very interesting problem. On May fifteenth, about ten o'clock in the morning, I was out in our corn patch hoeing weeds when I looked down and saw a tiny cluster of bees among the clods. I got down on hands and knees and, sure enough, there was a beautiful Italian queen bee with an escort of thirty workers! A glance showed that she was young, healthy, and fertilized, ready to lay eggs if she just had a home. But where was the swarm that should have accompanied her? There was none. I hurriedly got a small wooden box and placed the open side near her so that the sun warmed the interior. She went right in and her escort with her. When I told my father, he looked at them, shook his head, and laughed.

"You'll never make a hive out of that half a handful of bees," he said.

"Don't be too sure," I told him. "Let's experiment. Let's hive them late this afternoon."

To our surprise, as the hours passed, a few more bees, two or three at a time, kept coming to join our tiny swarm in the box. At four thirty that afternoon I made as careful a count as I could and found that another 300 bees had come to join our queen and her escort. Now we had a grand total of 330 workers and a charming queen bee. By weight we had approximately three-quarters ounce of bees, a far cry from the two or three-pound packages one may buy to start a hive, and a very far cry from our usual six to ten pound wild swarms.

We hived this little swarm in a super type hive with two drawn combs without honey, keeping out the other eight frames to leave room for my big pan feeder. Into this we poured a pint of honey. We closed down the entrance to a small opening at one end one inch long by five-eighths inch high. Three days later our new field force was hard at work carrying mostly pollen. We timed the returning bees at six per minute. Not bad, I thought. On their first play flight when the hive bees came out for air we counted a total of eight bees flying around at one time. My father again shook his head and voiced his previous negative prediction.

However, when we opened the hive ten days later, the pan was empty, with the honey neatly stored in the honeycombs; so we gave them another pint. Also, the queen had been busy laying eggs, for the hive bees were covering a patch between the combs about the size of a small man's hand. At the end of thirty days when the first new bees were added to the field force, we noticed a definite increase in activity around the hive. At forty days we counted twelve bees landing per minute during the heat of the day. At sixty days there were eighteen per minute, and at ninety days forty per minute, and at one hundred twenty days we reached our highest count of seventy-eight bees landing per minute loaded with nectar and pollen on a half and half basis. At this time we removed our pan feeder, having fed a total of five pints of honey. We took eight drawn and quite well filled frames from another hive and gave them to our tiny hive to supply them with honey for the winter. That fall on October second we estimated our experimental "Tiny" hive to have a total of 8,000 bees or almost two pounds, and I fervently hoped they would be able to survive the winter. Even my father gave them a sporting chance to win.

During a normal winter we expect to lose one hive out of every five we own. This has been our average over a period of twenty years. More important than just experimenting to see if we could pull the aforementioned small swarm through the winter was the safety factor it gave us for our other hives. If one of our big hives became queenless we could immediately introduce the queen of our tiny swarm and so save a possible world's record breaking hive from becoming a total loss.

What happened to "Tiny" that I found in the garden among the clods in the spring of 1972? It survived the first winter and increased steadily in numbers throughout the spring of 1973 and drew consid-

Chronology of Hive No 4

Supers Added & date

January 30	•	Queen Excluder & first super
February 20	•	Replaced old excluder with new one & added super
March 12	•	One Super
March 13	•	One Super
March 20	•	One Super
April 4	•	One Super
April 12	•	One Super
May 27	•	Two Supers
June 6	•	One Super
June 7	•	One Super
June 13	•	One Super
July 1	•	Two Supers
July 17	•	One Super

erable honeycomb but was unable to produce any excess honey that first season. However, it was evident to us that our lovely queen had unusual egg laying ability. In fact she kept on laying eggs, with resulting brood and young worker bees being added to her work force, all through the fall of 1973 and early winter of 1974. My father and I realized that with this hive we had a golden opportunity to set a new world's wild flower honey production record. With eager expectation I lavished upon our queen and her bees all the love, know-how, and expertise we had gained in fifty years of beekeeping. The result—a new world's record—404 pounds of delicious honey as of September first, 1974!

Many folk who came to visit us in 1972 and had an opportunity to observe "Tiny" have since asked us what became of that precious little hive. They often peek around the corner of our house to look at what they think is "Tiny" without saying anything to us, but are puzzled when they no longer find just a few ambitious little workers coming and going but instead the constant hum and zoom of hundreds of bees per minute returning to a vast multi-storied structure which, during the maximum honey flow this past summer, housed an estimated 135,000 bees in nine medium depth supers above the queen excluder. We then explain to our visitors that we found it necessary to rename "Tiny" as Hive Number Four due to its steady growth in 1973 and huge size this year. They all rejoice with us that this beloved little swarm has produced a new record.

19

Storing Empty Combs For The Winter

THE IDEAL WAY to store empty combs for the winter would be to extract them, have the bees clean them of all honey, and then put them away in a clean dustfree cool place. However, to attain this goal is something else again. We have tried several methods and none of them have been entirely satisfactory. It would seem that one could extract six or eight supers, remove the cover from a hive, and then stack all these empties upon the hive about September first, and their being no honey flow, the bees would clean them up in short order. Indeed, they do clean and polish the frames very quickly, but instead of carrying the honey down and storing it in just one or two supers they often neatly place the honey they have cleaned from each frame back into a few cells of that frame. This is no help to us as beekeepers because we want the frames completely emptied of all honey so that no honey will crystallize in the cells over winter or be an attraction for ants.

We have tried extracting ten supers, and around three o'clock in the afternoon of a warm day, have set them all outside stacked crisscross for all the bees in the neighborhood to work on and clean up. Believe me, this works, at least as far as the cleanup is concerned. In a short time there will be a tremendous buzzing with frantic activity all around the stacks of supers. But there are two great drawbacks to this method. First, in their frantic haste to get at the honey the bees climb over each other and tear up a lot of good wax comb. They act like vandals and robbers in their haste to snatch the best of the honey, and whereas in their own hives they are always careful and neat, now they sometimes rip, tear, and destroy comb with utter abandon. If we have set out old and tough, often-used combs, it is not so bad, but with combs newly drawn the past summer it is bedlam and destruction. The second disadvantage is that it frequently starts the bees to robbing from each other's hives. Having once tasted fresh, easily obtainable honey, they go on the prowl for more. No doubt they also remember the scent of the combs they have just cleaned, and when they detect that same scent in an established hive they make every effort to go in and rob. One cannot really blame the bees. Today we want them to clean up a stack of supers from various hives in our apiary, and tomorrow we condemn them for trying to continue the good work. We came to the conclusion that this type of robbing was our fault, not the bees'. Of late years we have abandoned this method.

What we do is to extract the combs as cleanly as we possibly can and stack them away. What honey remains will crystallize by spring and, when we put them on again for the bees to fill, the honey the bees store in them will tend to crystallize rapidly. This presents only a minor problem. Many people like honey in its crystallized form and we have sold half of our production in this form. If we do need more liquid honey than we have on hand we reliquefy some jars in the warmer.

Caution: When stacking extracted supers away for the winter, always lay a sheet of cardboard between each super and the next higher one. This helps greatly to reduce the damage done to the combs during the winter by the activities of the wax moth worms and to keep out ants.

If we do unknowingly have wax moth eggs in a super, the larvae or worms cannot easily go up or down and destroy the whole stack of supers. The cardboard between tends to confine them to the one super. Furniture stores often discard large cardboard cartons suitable for cutting into super covers. We use an empty super box for a pattern to mark out the cover size on the cardboard and then cut the cardboard with a linoleum knife. Even five sheets of newspaper between supers will help.

About October fifteenth we go through all our stacks of empty supers looking for wax moths or worms. If we see a honeycomb with what appears to be cobwebs clinging to it or running across the face of the comb, a close examination will show that a worm has passed through. With an ice pick we scrape out these runs and webs. A short search will usually reveal the worm which we remove. We damage some wax comb in cleaning out the runs, but the bees can repair what we damage more easily than what the moths destroy. If we absolutely cannot find the worm, we remove that frame from the super and place it in a refrigerator or freezer for twenty-four hours. After the cold has destroyed the moth larvae and eggs we replace the frame in the super. It is not that the worms are too small to be seen with the naked eye for they are from one-quarter inch long when young to a full three-quarters inch long just before they go into the pupa stage. But they are positively expert at hiding themselves in the comb or in a hole gnawed into the wood of a frame.

Before we pass judgment upon the wax moths as being useless destructive pests it is only fair to say that they sometimes do us a great deal of good. For instance, if a swarm of bees dies out of a hive, hollow tree, wall, or chimney due to disease or other cause, the wax moths move right in, eat the wax, and destroy and pollute everything in the hive or cavity so that no other swarm can take possession. This is a blessing for us as beekeepers because we want each of our hives healthy and strong. In situations where the bee inspector cannot possibly make an inspection the wax moths take over his duties.

Caution: All year long, always be sure that all wax-filled frames are put away in supers or hive bodies tightly closed with bottom and top of some kind before dark every evening. The adult wax moths fly and lay their eggs mostly during and after dusk.

It always amazes me how rapidly the wax moths can destroy unattended or improperly guarded honeycombs. Years ago on June first, I noticed some bees flying around checking out each knothole and crack in our barn. This is a first indication that a hive somewhere close by is almost ready to swarm. These bees are scouts looking for a new home. Sometimes they come as much as three weeks in advance of the actual swarm. As soon as I saw these scouts, I set out an empty hive with all new, fully drawn combs. I carefully examined each frame for wax moths, eggs, or worms but found none. Almost immediately the scout bees found my hive and a dozen or so took possession of it. Ordinarily I would have closed such a hive just before dusk with a solid full width entrance cleat to keep out the wax moths. However, I had not seen any moths flying around as yet, and since some of the scout bees seemed to want to stay in the hive overnight, I left it open. There is a tricky point of procedure involved here. If scout bees stay overnight and we lock them in, they rarely bring the main swarm as we so much desire them to do. But if we do not close the entrance every evening we risk invasion by the wax moths.

Every day for two weeks, more and more scout bees visited the hive until over one hundred had it in possession during the daytime and about fifty stayed every night. I was truly pleased for it looked as though a large swarm was coming very soon. Days passed but the swarm did not come. By the end of the third week there were noticeably fewer scout bees around the hive. Disappointed and puzzled, I left the hive as it was for another week. Then I took off the cover and looked in. The whole interior of the hive was a complete and absolute mess! No wonder the new swarm had not taken possession—there was nothing for them to possess. All the beautiful wax honeycombs had disappeared from the frames. There was not a vestige of wax left anywhere. It had all been eaten by the wax moth worms, and these worms were everywhere. I counted eighty-five huge fat worms in their webs clinging to the cover board. The whole hive was filled with the webs and manure of countless others. Some had even bored holes into the new pine frames as they prepared to go into their pupa stage. In dismay and disgust, I built a roaring fire in our incinerator, pried out the ten foul and fuzzy frames and dumped the whole mess into the fire. In exactly thirty days the interior of my fine new hive was reduced to a total loss. Did I give up? No, for there had to be some way to encourage scout bees to stay for the night and at the same time keep out the wax moths. Catcher hives as described earlier have proven to be our best answer, as yet, to the problem. Wax worms will not, as a rule, eat refined wax starter. They relish wax with a little honey added as a sweetener.

20

Sweet Scented Beeswax

Beeswax, a by-product of the production of honey, has value. Most of the wax comes from the cappings which are cut off the sealed combs just before extracting. As cappings are mixed with considerable honey, we remove them from our large uncapping pan and into a large colander over a container so that the honey can drain out of them. Some beekeepers run these cappings through the extractor to remove the honey but when one has only a few hives it seems easier to let it drain out naturally. Cappings are very nearly pure yellow beeswax and care should be taken while uncapping frames, not to mix the bee glue (propolis) frame cleanings with the cappings.

Propolis, a substance manufactured by bees from plant resins, is dark brown or reddish in color and is quite hard. It is found in abundance on and between the ends of the frames in each hive or super as well as in many other places in the hive. The bees use it to hold everything solid and secure as well as airtight and waterproof. I have seen old abandoned beehives in such a dilapidated condition that it would seem impossible for bees to survive in them. A closer look showed bees still hard at work. A careful examination revealed that the old cracked and broken boards had been marvelously patched together with propolis. It is truly remarkable how much skill and ingenuity our friends the honeybees exhibit in solving their problems.

After the cappings have been in the colander for approximately two days, all of the free honey has drained out, but some excellent honey always remains. We store these cappings in a large covered container. When the container is finally filled, we scoop some out and fill another one-gallon colander with these now somewhat dried cappings and place it over another container to again catch the honey. We place a small electric plate upside down as a cover over both the colander and container and turn on the power, being sure the wires of the electric plate do not touch the cappings. In a short time the warmed honey remaining with the cappings runs down into the container below. Soon the wax cappings also begin to melt and run down. In about twenty minutes all the wax is melted so we shut off the power and let the honey and wax cool. We find a quantity of good honey below with a layer of semirefined wax hardened to a cake on top. As this honey has been heated to some extent, we cannot sell it as raw honey, but many people like it and use it in cooking. It is always very good thick honey.

Whenever I have accumulated six or eight cakes of wax, as described above, I give them a second refining. This can be done in several ways. I take a kettle of two-gallon capacity and pour one gallon of water into it to take up the honey remaining in the wax as the wax melts. I break up each wax cake into several pieces and fill the kettle level full. This kettle is then placed on an electric hot plate, or on the gas burner of our stove if I wish to work in the kitchen, and is heated until the wax begins to melt. The melting process must be watched closely, for if the burner is turned well up the wax may melt faster than expected. The water and wax should come almost to a boil but not actually boil. Overheating causes wax to become darker in color with correspondingly less value. If it seems the wax is going to boil before it is all melted, the heat must be turned down a little. I use a long spoon or flat stick to keep the mixture in constant motion. Stirring gives me an indication of how much wax still remains to be melted as I can feel the spoon bump against the submerged lumps of unmelted wax. Stirring also helps to keep the entire mixture at a more even temperature. If the wax begins to boil, an almost immediate formation of foam will gather on top. This foam will build up with astonishing speed and overflow the kettle if the heat is not instantly turned down.

Caution: Overflowing foaming wax burns readily so extreme care must be taken not to burn down the house!

I always carry the hot kettle of melted wax outside and pour it into forms set out in advance. They may be clean lard pails, plastic ice cream containers, the bottom part of a plastic bleach bottle or almost anything handy that will hold hot wax. If the container will hold boiling water it is safe to use as a form for hot wax. As a rule wax sticks tightly to the sides of any form unless we pour a few cups of boiling water into the form first and then pour the melted wax into the hot water. Thus as the wax cools and shrinks in size, it pulls away from the sides of the form and makes a lovely smooth cake. I have been told that forms rinsed with soapy water will also keep the wax from sticking, but when I sell pure beeswax I want it to be pure—without even a trace of soap or anything else.

Before pouring the melted wax into the forms, I always have ready a large (at least eight inches in diameter) wire strainer with a handle. Across this I place an eighteen inch square of closely woven nylon material with a much finer mesh than that used earlier to strain honey as it came from the extractor. I carefully push the nylon down all around in the strainer and hold it in place with clothespins so that the wire supports the nylon. I pour the hot wax through this nylon filter into the form and when it is full, shift the filter over to the next form. This way I can fill all forms rapidly and safely and at the same time strain out almost all of the remaining impurities. A little paddle should be at hand to squeeze the residue in the strainer, enabling one to salvage as much wax as possible. I discard this residue and, when the filter is cool, rub the nylon clean between my hands for use the next time. On the under side of the cooled cakes of yellow wax there will usually be found some gray material. This is wax combined with air and water. To make the cake look better I always take a knife and cut off this residue until the clean yellow wax appears. I save this gray wax and melt it again with the next batch of cappings.

Refining wax takes time and effort. On occasion we have lived in an area where there was little call for pure beeswax. In that case we melted it down into cakes as given in the first step in refining and stored it carefully away in five gallon cans with tight covers to keep out the wax moths. In a few years, after we had accumulated a quantity of wax, we took it to Chico, California, to the Diamond International Corporation, Apiary Division where they bought it for sixty cents a pound. They buy nice yellow cakes and also old black crushed up combs and slumgum (old, brittle, useless combs) of any kind. They have the facilities to extract and refine the wax from all grades of beeswax even if it contains propolis or foreign matter. They grade the material brought and pay accordingly. If we have accumulated at least one hundred pounds of good wax they will also refine it for us and make it up into any type of new starter comb that we need for fifty cents a pound. This price may vary from year to year as well as company policy so we always write a letter of inquiry first to see if there has been any change in price or policy.

People use wax in many ways—for batik work, candles, wax sculpture, furniture polish, making bullets for reloading cartridges (one teaspoonful of wax per fifty bullets of melted lead wheel weights to compensate for the antimony), wax modeling, and other uses. If we have honey to sell we are expected to have wax also. We make beeswax into cakes weighing from two ounces to three pounds so that our customers may have a choice of size and color. Color varies from brown to almost white. Most people want beautiful yellow wax. For years we have sold refined wax for an average price of ten cents an ounce.

As beeswax has a delightful scent, I often make special little cakes just for the fun of it. These I give away free to those who come to us for honey and who express a desire for a little wax. It helps folks to remember us the next time they need honey as well as retain pleasant memories of their visit.

21

Honey For Health & Profit

HONEY IS A GREAT natural food that goes largely to waste in this country and I suppose in every other country of the world as well. Bees and flowers are to be found in all parts of our land. Yet the flowers, and flowering trees and shrubs, so outnumber the bees that only a small fraction of the total honey crop available is actually gathered. What a blessing it would be to mankind if more people would learn to hive and care for bees. Honey is so delicious and so good for our health and well-being that we should give more attention to harvesting this almost universal source of food and energy. Through the years I have talked to many aged people and almost all of them have been users of honey. One of my long-time honey customers told me recently that she will be ninety-eight years of age this coming July. She is so active and cheerful that I would never have dreamed that she was that far up in years.

"God has surely been good to you!" I exclaimed when she told me.

"Yes, He has, but I've also been good to myself," she answered. "Bring me some more honey as soon as you get some for I use no sugar and my honey supply is running low."

I assured her that I would surely remember her. I am very pleased to see that recently more and more young people are also beginning to use honey. I talk to many young folk who now use honey in their coffee, tea, breakfast foods, and home canning projects. For our own use my father and I regularly eat a teaspoonful or more of honey with every meal. The amount of honey the body can readily assimilate depends to a considerable degree upon one's daily physical activity. Strenuous physical labor requires a greater intake of concentrated food, such as honey, than does a more sedentary mode of life.

Sometimes when working with our bees we do not have time to sit down to a regular meal. Then we make a meal of just honey. When we are hungry we can eat enough honey to satisfy our needs provided we have plenty of drinking water. We eat a couple of tablespoonfuls of honey and take a drink of water, repeating the act as often as necessary. Honey is such a concentrated food that to eat much of it without water can make one feel ill.

People who have diabetes know that they cannot safely eat honey any more than they can eat sugar. Yet through the years numbers of diabetics have bought honey from me, especially comb honey. They usually buy the biggest, heaviest, most beautiful comb I have available. One day I asked one of our elderly diabetics how he was able to eat so much honey and still maintain his present degree of health.

"Eat it!" he exclaimed in astonishment. "I don't eat it. I just look at it. I never take it out of its transparent plastic wrapper. I set it on a little shelf above our breakfast table and enjoy its beauty. I have one there now that I bought from you two years ago and I want this one to place beside it for comparison purposes. The first comb now has a brilliant white color. This honey brings back happy memories of my boyhood buddies." And a big smile crossed his face.

I realized that he had just told me one of the secrets of joyous living. In my case, I may not drink tea or coffee but that need not stop me from enjoying their pleasing aroma. So it is with honeybees. If we live where we cannot have bees of our own we may yet enjoy watching these marvelous little creatures, for during the warm days of spring and summer we may observe them busily at work on the blossoms along almost every roadside.

Bees make honey in a wide range of colors, consistencies, and flavors—from almost white to dark brown, from slightly more viscid than water to so thick it will hardly flow, from exceptionally sweet to peculiarly bitter. The honey they produce depends entirely upon the nectars they find in any given locality the world around. A curious fact in our area is that during the last days of July and the early days of August, some of our bees bring home to their hives nectars that make pink honey, and others bring nectars that make a beautiful jade green honey. The bees fill a cell here and there on the comb completely full of pink honey or green honey. They never seem to mix them. These unusual honeys have a different but pleasing flavor. I wonder how many other beekeepers have observed this phenomenon.

We should not criticize the bees for sometimes making honey that we do not especially enjoy. We must remember that the bees make the honey for their own use, not ours. If we can encourage them to make an additional amount for us, we should be thankful. Sometimes, however, we can do much to determine the flavor of the honey the bees make as surplus for us. Here in the lowlands along Monterey Bay, California, our wild flower honey has an excellent flavor all season long. But in the mountains above us only the early spring bloom makes really fine flavored honey. We watch carefully, therefore, and as soon as this early honey is sealed we remove and extract it. We give the empty combs back to the bees to refill with whatever good or peculiar tasting nectars they can find in the latter part of the season to store as honey for their own needs. Of course we always make certain there is still a nectar flow so that our bees do not go hungry. Thus we harvest some fancy grade honey from an otherwise unprofitable area.

Even though we have set a new world's record, we may not be able to keep a sufficient number of hives in our present location to yield as great a quantity of honey next year. Several large apartment-complex projects are scheduled for construction in this area in the near future. It seems now that we will have to look for another location from which to top our own best year of production. It is of no use trying to stop the wheels of commerce and progress; far better to pack up and go on one's way rejoicing. Another record-breaking location may be just over the next range of hills, or many miles away.

When we plan to move a considerable distance we do not take our bees with us but sell them and begin again. We do this by catching wild swarms in our new location. The reason is that bees, like human beings, need to become acclimated and this takes time. I know that bees are sent or shipped thousands of miles and this is all right for they will gather good quantities of honey and do an efficient job of pollination. However, for our purpose, my father and I have found local bees to be better maximum producers.

Someone might well ask, "Have any of the people to whom you have sold bees really made a success of it?" My answer is a resounding, "Yes indeed!"

Ten years ago a man bought three hives from me. Recently his wife met me on the street and exclaimed, "You are the honey man!" I acknowledged the fact and she introduced herself. I then asked her if they had obtained any honey from the bees that I had sold them.

"My, I should say so," she replied. "We have so much it almost runs out of our ears!"

Then she told me that at first they had but little success. They attributed their near failure to having their bees in the wrong location, so they moved them several times. They first moved their hives to a place in the middle of a large windswept field. Next they moved them to a location out of the wind but under some large shade trees. In each instance the bees survived but produced no surplus honey. Now for some years they had set their hives in a warm natural clearing at the edge of some woods and their bees had done wonderfully well for them. From time to time others also tell me of their successes.

So in answer to the often asked question, "Do you think that I could learn to keep bees?"

My answer always, "I should think so. Why not give it a try? Fan your spark of bee fever into a bright shining flame and give it a try!"

God is Love and Love can do miracles even with bees!

22

Addendum: A Bee Bachelor's Domestic Dilemma

When we celebrated my forty-fifth birthday and my mother and father their seventieth, my mother said to me, "Ormond, let me teach you how to cook. Some day you will need to know." Near panic seized me. I like to eat, but cook—oh no! In the years that followed Mother made numerous attempts to introduce me to what she called the pleasant and challenging intricacies of the culinary art, but to no avail. The great outdoors and our beloved honeybees filled me with a call of the wild that I could not resist.

After Mother passed away, my father became cook. He had often related how during World War II, when Mother had taken a trip to Los Angeles, he had done his own cooking for six weeks and worked every day besides. It was a good story as far as it went, and it went just six weeks with me. We always had two boiled eggs for breakfast and that was about the sum of it. One morning I ventured to voice a mild protest. I decided that a bit of descriptive language might help the situation, so I said to him, "Father, these eggs have been boiled so long they are positively blue and for texture and density would pass for gem quality lapis lazuli."

He answered softly, "Sonny boy, do I understand that you object to my cooking?"

"That I do!" I replied.

A soft smile touched his lips. "Now let's get this straight," he said.

"The primeval law of the masculine jungle states clearly, quote, 'He who does not appreciate the manner in which the food is prepared must himself do the service of cook until some other rash unfortunate raises an objection,' unquote. Do I make myself clear? You are now chief cook! Congratulations my dear boy."

The primeval law of the masculine jungle! Memory rushed back through the years. Somewhere I had heard of that law but never in such a fancy disguise. Ah, yes, now I remembered. Men spoke of it as "put up or shut up." Well, if my father thought that a little descriptive language on his part would get the idea across more clearly to me, he was right. We continued our breakfast in silence.

As we each polished off our last egg I had a minute to think things over. I had opened my big mouth and Father had stuffed an elephant's foot in it. I watched my father closely. As long as he was cook, he washed the dishes and I dried. As usual he packed the dishes into the wash pan, and ran the hot water. Then he stepped suggestively aside. I caught his action out of the corner of one eye. Wheeling around to the sink, I picked up the wash cloth. Our eyes met and we both laughed. I was now officially installed as cook.

We both went out to work with our bees but since it was a rather cold cloudy day in late winter there was not too much to be done. Father suggested that I go back to the house and tackle my new job as cook. I agreed. Soon I discovered why we were on such a limited diet. There was almost nothing in the house for a working man to eat except several dozen jars of assorted canned fruits. A box of apples caught my attention. Visions of a big apple pie floated vividly before my mind's eye. I would make an apple pie! Hurriedly I sorted the apples in the box. Many of them were rotting but I still salvaged a water pail full of good ones. Years ago I had made an apple pie under the urgent and skillful direction of my mother and it had turned out well. For an hour I peeled and cored apples. Just as I was happily finishing the last one my father came in.

"What in the world are you going to do with all those apples?" he asked.

"Make an apple pie."

"How many pies?"

"One, of course."

"With a gallon and a half of apples? Must be a giant-size pie you're planning?"

Just then someone knocked on the back door. My father turned to answer it. I listened. It was what I had expected, someone to buy honey. My father had a loquacious customer. I turned to my pie baking again. In dismay I looked at a gallon and a half of prepared apples. I had been so busy trying to peel and save all the good ones in the water bucket that I had forgotten to calculate the number it would take to make a pie and, to be truthful, my thoughts had all been out with our bees working on a knotty problem of bee culture.

At least there was one tiny bit of silver lining showing—my father was occupied elsewhere. What was that remark of his about a giant-sized pie? The answer, of course! Once in the long ago I had helped Mother make a pie and I had peeled too many apples that time too. In reply to my query, "What do we do now?" she had smiled and answered, "We make a bigger pie." Then I had asked her, "How big a pie could we make if we made the biggest pie we could make?"

"Well let me think a moment," she had replied. "Somewhere in this kitchen we have a pan twenty inches long, sixteen inches wide, and two inches deep, and that is how big a pie we could make."

What a pie pan! That was the one I needed right now. I glanced out the back door. Father was still talking. Good. I did not want him to catch me flying around the kitchen opening drawers and doors searching for an elusive pie pan. But where was it, anyway? I paused for breath in the middle of the kitchen. A glint of light reflected from some object in the cranny between the refrigerator and the wall. My lost pie pan? Sure enough, it was. Hastily I washed and dried it, then poured in the apples. A sudden thought stopped me cold. Pies have crusts! So they do. Slowly I poured the apples out of the pie pan back into a big kettle. In my frantic search for the big pie pan I had found the sugar, flour, shortening—in fact all the ingredients for a pie crust. I had even found a big mixing bowl, rolling pin and rolling board. I set everything in a row on the work table as I tried desperately to remember how a pie crust should be made. It was no use. I had not the slightest remembrance.

"Oh, Lord!" I cried aloud. "What do I do now?"

Then I heard it—the Voice! It said, "Four cups flour, one-quarter cup sugar—." I measured out and mixed the ingredients at top speed. As I divided the big lump of dough into four parts, two for the bottom of the pan and two for the top cover, and began to roll them out I recalled other occasions when I had heard and followed the directions of the Voice.

One spring day when we had gone for a drive up the coast highway, in the vicinity of a little town called Davenport, we stopped to stretch our legs. I had walked on a road along a high embankment that dropped off to the left. On looking over the edge I saw something away down at the bottom. The Voice said, "Go down to the chair." That was quite an order. I could slip and slide down but getting up again would be another matter. I almost disobeyed but thought better of it and started down.

Just then my mother called out, "Where are you going?"

"Down to the bottom."

"Why?"

"Because I have to."

She shook her head.

Down at the bottom there actually was an old overstuffed chair. It had been there a long time. The fabric on the arms was completely rotted through. I pressed down on what was left of the cushion. Then I saw it—a purse—wedged between the cushion and one arm of the chair. I pressed down harder and retrieved it. Old, moldy, and rotten, it cracked open from the pressure of my hands. There was money in it! Down there among the weeds and bushes was no place to examine such a treasure. I tucked it carefully in my jacket pocket and then scrambled for the top of the embankment. My father reached down a helping hand as I came to the slippery shale near the top. Breathlessly, I handed the old purse to my mother.

"Let's take it to the car," she exclaimed softly, "and lay it on the fender."

Slowly and carefully she pulled the old purse apart. Age and weather had almost solidified the contents. Deftly she separated a twenty dollar bill from the mass, then another, and another, plus two one dollar bills. Sixty-two dollars in all! Her eyes shone with excitement.

"What made you go down there?" she whispered in my ear.

"The Voice," I whispered back. She nodded.

"Let's go home," she urged. "We've had enough adventure for one day. Since there is no sign of identification, the money is ours. We must wash it carefully in soap and water to remove the mold and musty smell."

As I rolled out the fourth lump of dough to finish the top crust another encounter with the Voice crossed my mind. I had stopped at a rummage sale and was looking over an extensive array of exquisite junk when my eye caught sight of a large cigar box on the floor.

"Buy it," whispered the Voice.

I picked it up and opened the lid. It was full of small buttons. Buttons of all things! I scratched my head in perplexity. But the order was to buy it so rather hesitantly I made my way over to an elderly lady who was running the sale.

"What do you want for this box?" I asked timidly.

"Twenty-five cents," she replied.

"I'll take it," I said, and handed her the money.

"Your wife must do a lot of sewing?" she ventured.

"No, she doesn't."

"Then tell her to learn," the lady spoke rather sharply.

"I can't do that."

"And why not?"

"I'm a bachelor," I answered meekly.

Her lips tightened into straight lines as she drew a deep breath. But before she could articulate a vociferous admonition I had scooted around a corner of the house. No use waiting around for her to tell me that a wastrel and his money are soon parted. My ego was in very short supply at the moment as it was.

As I drove slowly home I wondered if I should tell my mother. She was not able to be out much but was always interested in my doings. I parked my car and walked with lagging steps toward my mother who was sitting on the porch.

"You have been to a rummage sale," she said.

"Yes, but how did you know?" I asked in suprise.

"You bought something and now you are almost afraid to tell Mother?" she inquired confidentially.

"Right again," I answered more hopefully.

"What did you buy?"

"A box of buttons for twenty-five cents." I went back to the car and got them.

"Why?"

"The Voice."

"Let's go right into the house and pour them out on the table. Then we can see what you got for your money."

As we sorted through the buttons Mother found a penny. Then I found two more and Mother another two more. Five old pennies!

"Well," said Mother, "you could have done worse. You are only out twenty cents. I see no old or rare buttons."

The old pennies interested me. I picked up our biggest enlarging glass and examined them. They had all been minted in the 1920s. Four were United States coins. The fifth was a 1921 Canadian penny.

"I wonder if this Canadian penny has value?" I spoke out loud.

"It does," whispered the Voice.

"Mother, it is the Canadian penny that has value. I will show it to a coin dealer when I go downtown tomorrow."

Next day when I saw the coin dealer he offered me three dollars fifty cents for it. I was about to sell on the happy spur of the moment when my eyes caught sight of the dog-eared condition of the reference book he was using. All his other coin books looked sharp and new. I decided not to sell. He repeated his magnificent offer. I said nothing but held fast to my penny and went home.

"What do you think he offered me?" I asked Mother.

"Twenty cents, maybe? Then you would break even on your button deal."

"Three dollars and fifty cents," I told her happily.

"Let me see the money!"

"I didn't sell, would you have?"

She heaved a great sigh and slightly shrugged her stooped shoulders. "You take after your Grandmother," she answered. "She did some queer things, too."

However, a short time later I sold the penny to another coin dealer for eight dollars and Mother forgave me. The twinkle returned to her eyes.

The Voice. What a blessing it had been to me through the years and was again in my pie baking dilemma. My oversized pie was almost ready for the oven. As I began to pinch dimples all around the edge of the crust my father rushed into the house.

"Ormond, hurry up with that pie," he almost shouted. "Our experimental hive No. 6 has a large mass of bees on the landing board!"

In feverish haste I finished pinching the dimples, poked a few holes in the crust, shot the pie into the oven, turned the heat to 350 degrees, and tore outside. Our experimental hive No. 6 was one of four hives in the apiary that we had high hopes would set a new world's wild honey production record from one hive in one season. We had to quickly solve the mystery of the mass of bees on the landing board or our chances of a new world's record from that hive would be gone forever.

I lost all track of time. The pie in the oven was completely forgotten. My thoughts were all centered around our beehive. Dimly I heard the Voice whisper, "Turn off the oven."

"Turn off the what?"

My pie! I ran to Father and screamed into his somewhat deaf ear, "My pie!" He looked up startled, then waved me toward the house.

I took an apprehensive peek into the oven. The pie was beautifully browned and done! I shut off the oven and took out our pie. Obviously this pie belonged much more to the Voice than to me. Again I thanked the Lord of Glory for His loan of the Voice.

I suppose it was inevitable that the day would come when I would disobey the Voice. We were building beehives out in our shop and I was busy sawing out parts for my father to nail together. He was away at the grocery store shopping. He always did the shopping and bought whatever pleased him. It was my job as cook to turn his purchases into something tasty, or at least palatable, or at the very least capable of maintaining soul and body as an entity. Too often it was the latter. I could not be proud of my cooking.

As a rule my father came home with a frying chicken. We had boiled frying chicken for several months. I had tried several times to fry a chicken but without success. The under side always smoked and burned before the meat was cooked through. To stop the smoking I poured in a cup of water, put on a tight lid, and we ended up with boiled chicken again. At supper one night my father asked quietly if it would be possible to have fried chicken some day soon.

"No need to wait," I told him. "We have fried chicken tonight. Just open the pot and dig in."

He took off the lid with alacrity. Then a puzzled expression spread over his face.

"It's been fried all right," I assured him. "I fried this fowl until its bottom turned black. Then I poured in some water—" Father dug out a drumstick. "Quite tasty," he said. "Really, you are improving."

I tried a piece myself. It did have a different flavor from what we had been having. I was not sure whether it was for better or worse, but at least it was different.

This day he came home with a two-pound chunk of beef.

"Got any ideas on how to fix this beef?" I asked.

"Yes," he replied. "Cut it into two pieces and put it in the usual pot with a little water. Not too much water, though. Then come out and let's get to work."

The meat looked tough. It would soon be lunch time. I turned the burner a little higher than usual, and went out to the shop. We worked away for some time. Then the Voice whispered, "Meat's done."

"Good, but I have just a few more boards to cut out." So the time slipped by and the meat was forgotten. Much later my father looked out of the shop.

"I'm hungry," he said. "Go get us some lunch."

As I neared the house I saw smoke coming out of the kitchen window. Strange! Was the house on fire? I dashed into the kitchen but could see nothing. The whole house was full of greasy blue smoke!

"My meat!" I gasped. I groped my way to the stove, shut off the gas, and ran out the back door with my kettle of meat. I took off the lid and looked in. What I saw was even more depressing than I had expected. The two pounds of beef were now two little black balls the size of hen's eggs. I sat down on a bench and just sighed. Where was the silver lining to this black cloud?

In the distance I heard my father slam the shop door. He was coming home to lunch. Into the midst of my mental blackness there came a tiny ray of light. Along with the light there came an idea. Why not put the lid back on the pot and replace it on the stove? Maybe I could trick him into becoming cook again. Not that I really wanted to shove the job of cook back onto his shoulders for at age eighty he was already doing all he could. But it would be fun from time to time to remind him that I was doing his job. He was both tired and hungry. Mild-tempered as he always was, surely this time he would blow up and I could apply his "primeval law of the masculine jungle," routine and trap him. I felt almost cheerful.

"I smell smoke!" he said as he sniffed here and there.

"So do I," I replied.

"Fire in the house?"

"Come in and see."

The house had begun to air out. I lifted the lid to our smoking burned out meat pot. He peered in. I held my breath!

"By the great horn honey-spoon of my great uncle Adolphus!" he ejaculated in amazement.

"That's our two pounds of beef. Do you think we can eat that stuff?" I asked quickly.

He peered in again. "Well, now, let me see. If we were hungry enough it might be right tasty," he answered as he turned to give me a sly peek out of the corner of one eye.

I plunked limply down on the nearest chair. Father faced me with a sympathetic smile of understanding. "It's pretty hard to poke an old racoon out of his own hollow tree," he remarked.

"True enough," I admitted. "Utterly impossible."

Then the ludicrous aspect of the situation struck us and we both laughed. I was still officially cook. Thank God for our honey-bees—we still had some bread and plenty of honey.

Something had to be done about breakfast. Two eggs apiece plus a cup of hot water and maybe a slice of bread spread with honey were just not enough to sustain us until lunch time during the long summer days ahead. In one of the cupboards I had many times seen a large round box of Quaker Oats, the quick kind. Now I took it down and read the instructions. They seemed simple so I decided we would have a change of diet for breakfast next morning. My father always went out each morning for a first quick look at the bees while I prepared breakfast. When he came back in to the kitchen next day I had everything ready—one boiled egg and a big bowl of oatmeal mush for him and another setting of the same for me.

"Say," my father nodded approvingly, "this mush really hits the spot. Do it again." I did, and for many days thereafter.

Then one day we received a letter from my sister LaVerne who lives in Alberta, Canada. "We have just sold our machinery business," she wrote. "Would you fellows care to have me come down for a visit? I could stay six weeks."

My father was delighted. "Let's tell her to come right away," he beamed. I had a few mental reservations. I would be delighted to see her too if she would do the cooking. But the few times that she had visited us during the past thirty years, Mother had always prepared all of the meals. I remembered hearing my sister say that she could cook, but would she? I gave the matter a good think but could come to no conclusion.

"What have you got to lose?" whispered the Voice.

I thought that one over.

"Nothing, really," I answered. "Thank you for the tip." So I told my father to write to LaVerne and tell her to come as soon as she possibly could. We would both be charmed to see her. Really, I had everything to gain and nothing to lose. I began to look forward to her coming.

My sister arrived on the eleven o'clock bus one Wednesday morning. I happened to be in the house when she telephoned so I went right down in my car and got her. She was as vivacious as ever. My father had the table laid for lunch when we got home. It did not take me long to get our boiled fryer warmed up. I always prepared enough to last us for two days and this was the second day. Father and I ate as hungry men will. My sister tried a tiny piece of chicken.

"What do you think of my cooking?" I asked.

"Not bad," she said, and went on with her lively conversation with Father.

Came supper time. There was no move on her part to prepare anything so I did as we had always done—I put the same old pot on the stove to warm up its remaining contents for supper. Father and I always ate what there was and were thankful. My sister said she had been eating too much on the plane to be hungry. Could be, I had never been on a big plane.

When bedtime came I announced that breakfast was always at seven o'clock sharp.

"I'll be right there," my sister agreed.

Next morning I prepared three boiled eggs and three big bowls of oatmeal mush. I placed an egg in each plate and a bowl of mush alongside. My sister arrived on time and we sat down to breakfast. We said grace as usual and Father and I began to eat. My sister just sat still and said nothing.

"Your egg is getting cold," I dared to suggest.

"I don't eat eggs for breakfast," she replied. "I don't eat mush either."

My heart skipped a beat, then jumped with hope. The primeval law of the masculine jungle stated plainly in part, quote, "—until some other rash unfortunate shall raise an objection," unquote. Had she broken this cardinal rule of life? I analyzed her answer. She had just stated a simple fact that she did not eat eggs or mush for breakfast. She had not really criticized my cooking. I had to think quickly and calmly. I picked up my egg. It danced lightly in my fingers. Just a little bit of a push and I could slide her over the edge into my cooking job!

I had it! I would tell her—. But just then another thought intruded upon my bright idea. My egg stopped its dancing and my heart slowed down, then quietly sank without a ripple into the well of despair. This miserable intruding thought had substance to it, namely, that the primeval law of the masculine jungle *might* be different from the primeval law of the feminine jungle. As a bachelor of fifty-six years how could I possibly know feminine law, primeval or otherwise? My case was lost before the judge even made his opening statement. What chance did I have to present my meager evidence to the jury? None!

As I began to peel my egg, I said to my sister, "Dearie, if there is anything at all in this house that you would like for breakfast, do please help yourself."

"Thank you," she returned brightly. "I'll do that. One of those big red apples in the fruit bowl would do nicely for a starter, and you have a few other things around here that might tempt me."

The crisis had passed, but I was still cook. What was tragically worse, I was without hope.

After our usual devotions and prayer time, Father and I went out to work. LaVerne stayed in the house. About nine o'clock she came out to where we were and inquired matter-of-factly, "Would you boys care to have me do the cooking while I am down here?"

A sledge hammer blow between the eyes could not have stunned me more! Father gave a joyful instantaneous assent. I managed a feeble nod as a big lump arose in my throat.

"I rather thought so," she answered serenely. "This bee barracks is practically devoid of food. Who will take me shopping?"

"Father will," I gasped, as the lump dissolved in my throat and I got my breath and voice back in operating condition. "He'll be just overjoyed to take you, bless your lovin' heart!"

My sister laughed a gay and merry laugh. "For a minute there I thought you had seen a ghost."

"For a minute I thought I had heard one," I replied.

"Cooking is a lot of fun," she said, "but you are almost too late to learn. But take heart, keep trying, you may make it yet."

During the happy days that followed we feasted like kings. We had real fried chicken, fried potatoes and onions, corn on the cob from our own garden, pies and puddings. . . . The days sped by so rapidly they seemed to tumble one over the other. In addition to doing the cooking LaVerne answered the telephone, took orders for honey, sold honey to the customers who came to see us, and in general helped in every way she could. In what spare time we had she undertook to teach me to cook. At the end of each effort she would say, "Not bad, Ormie, not bad." But never once could I fix something that she judged, "Good."

We were indeed sorry to see my sister leave again at the end of six weeks. I really tried to put into practice what she had taught me. My father said I was improving and I really thought so too, at least I hoped so. Months went by and it was the fall of the year again. One evening as I was talking on the telephone I heard a knock on the back door. I motioned for my father to go answer it. Minutes later I heard him talking to a neighbor lady in the kitchen. Suddenly I heard her say, "What a delicious aroma coming from that pot on the stove!"

Could I believe my ears? I guess Father could not believe his either for I heard him ask her to repeat her last statement.

"What are you cooking for supper?" she asked. "It smells positively delicious!"

I did not catch my father's answer. My head was in the clouds for my ears had just been filled with the sweet music of the stars.

"Did I really hear her aright?" I asked aloud when our telephone conversation was finished.

"You did," whispered the Voice, "be glad."

"I am glad!" I whispered back in astonished reply.

I had attained the first pinnacle of success in the culinary art. A lady, bless her dear heart, had complimented me on my cooking. Or was it the sweet scent of fresh honey in the house that had helped to win her plaudits? No matter, either way her kind words were of monumental encouragement. With a bee fever stricken beekeeper it is love me—love my bees.

Index